ALLE ZEIT WACH
1842

G. Grupe A. N. Garland (Eds.)

Histology of Ancient Human Bone: Methods and Diagnosis

Proceedings of the "Palaeohistology Workshop" held from 3–5 October 1990 at Göttingen

With 83 Figures

Springer-Verlag
Berlin Heidelberg New York London Paris
Tokyo Hong Kong Barcelona Budapest

Professor Dr. GISELA GRUPE
Institut für Anthropologie und Humangenetik
Universität München
Richard-Wagner-Straße 10
W-8000 München 2, FRG

Dr. A. NEIL GARLAND
90 Conyngham Road
Manchester M14 5EA
United Kingdom

Supported by the Stiftung Volkswagenwerk

ISBN-13: 978-3-642-77003-6 e-ISBN-13: 978-3-642-77001-2
DOI: 10.1007/ 978-3-642-77001-2

Softcover reprint of the hardcover 1st edition 1993

Typesetting: Camera ready by author
31/3145-5 4 3 2 1 0 – Printed on acid-free paper

Preface

The examination of excavated human bone finds is mainly the domain of anthropologists and forensic pathologists, the former working with ancient and historical specimens, the latter with modern finds. The methodological and diagnostic approaches to these skeletal finds are the same, regardless of the time of burial. For physical anthropology, bodily human relics are dealt with as historical resources which give clues to ancient population structure, population development, life-style and subsistence. They are thus able to help scientists understand the present state of human populations. The identification of the finds, whether species diagnosis or the evaluation of individual parameters such as sex, age at death, body size and shape, kinship and pathology follows the same procedure used by forensic pathologists, whose task is the identification of bodily relics in cases of crime, mass disaster and the like.

However, there are other disciplines which benefit from excavated bone finds. Anatomy gains insights into the morphological variability of the skeleton in time and place. The implications for modern physicians and pathologists are at least two-fold: pathological specimens are suitable to unravel the distribution of many diseases and the susceptibility of individuals to pathogens in pre-antibiotic populations. In addition to this epidemiological aspect, exhumed specimens often exhibit advanced states of bone disease which are no longer or only very rarely present in today's industrialized populations because of efficient surgical intervention and pharmacological treatment. In addition, there is a specialized discipline involved in the analysis of the fossilized state of bone preservation, which is largely concerned with vertebrate evolution: palaeontology. But whatever knowledge is derived from bone finds, the prerequisite in every case is a full and proper investigation.

For many questions, such as sex diagnosis, a purely macroscopic inspection is sufficient. For other problems, a more detailed investigation of the specimen is necessary to obtain as much information as possible. Today, with the help of modern technical facilities, one is capable of studying the most detailed properties of excavated bone including its crystal structure and protein or DNA preservation. But, many people who are specialists in these fields of research are often not capable of interpreting a thin section of bone properly. Microscopic inspection of bone, whether modern or ancient, is a sequential step between macromorphology and ultrastructure. It is thus astonishing that laboratories where bone is routinely thin sectioned are scarce, and even so, the wide variety of technical equipment for microscopy is often not applied. Histology is, indeed, a prerequisite for many valid diagnoses; beginning with species identification in instances of fragmentary finds, followed by an estimation of the age at death, and, in addition, a correct pathological diagnosis is frequently impossible without sectioning the bone under study.

Therefore, we had the idea of organizing a workshop dealing with ancient, and also, of course, with modern bone histology to fill this gap. The fact that every participant who was invited by us immediately accepted the invitation might show that there is, indeed, a need for discussion on this topic. The workshop was intended as an occasion where both the research potential and the practical uses of exhumed bone histology could be discussed.

The conference was finally held from October 3-5, 1990, in Göttingen, in the rooms of Carl Zeiss Inc. Besides the conference room, Zeiss provided several light microscope facilities, thus the paper sessions were accompanied by practical sessions where specific problems were directly discussed and addressed with the appropriate histological section. It was a very fruitful conference where specialists from a variety of involved disciplines came into close contact and exchanged their knowledge and ideas. Forensic specialists, anthropologists, pathologists, embryologists and palaeontologists from Great Britain, Germany, France, Hungary, the Netherlands, Italy and the USA took part in this meeting. We are greatly indebted to them as well as to the other participants who joined the workshop for their motivation and engagement.

The meeting strongly stressed the methodological aspects, i.e. the above-mentioned prerequisite for a full and proper specimen investigation. The workshop's proceedings summarize the discussions in such a way that the reader who wishes to become engaged in bone histology may gain enough information to do so. In addition, much of the current palaeohistological research is reviewed.

The papers presented were organized into four sections: methodological aspects of palaeohistology, identification, pathology and, finally, artefacts. Firstly, one needs some information on the circumstances under which human mineralised tissues can survive in the soil. Even more important is to make oneself clear as to the type of questions that should be answered by bone histology (Garland). One has to be aware of the fact that thin sectioning means a partial destruction of the find, which may raise complications in cases of rare and very valuable specimens. Thin sections are usually investigated by light microscopy with a variety of special equipment. The choice of microscope, method and the deviant behaviour of ancient bone compared to fresh bone regarding staining and embedding is discussed by Herrmann. These complications are brought about by the 'decomposition' of bones by chemical and physical and biological agents within the burial environment which lead to micromorphological artefacts (Grupe and Dreses-Werringloer). Bones which are not decomposed but rather fossilized are a unique source for comparative vertebrate micromorphology (De Ricqlès).

Following a definition of the problem, the choice of the method and evaluation of artefacts, the establishment of several individual parameters commences. The basic question, especially in the case of very fragmentary finds, is whether a bone is really human (Harsáyi). Age at death can be assessed microscopically both on interred bone (Uytterschaut) and cremated remains (Hummel and Schutkowski). These qualitative results should be further quantified by the advantages of microradiography (Heuck) and histomorphometry (Boivin and Meunier).

Many problems still exist with regard to the correct diagnosis of pathology, since no data on soft tissue or body fluids are available (Bianco). Pathological diagnoses on bone thin sections are demonstrated with two disease complexes: metabolic bone and joint disease (Boyce) and malnutrition and infection (Schultz). Wakely's paper on bone trauma is the only one which focusses on SEM studies rather than light microscopy.

It was felt that the advantages of SEM and TEM imaging of human bone also deserve much more attention than is currently paid. The broader discussion of these techniques, however, would have been beyond the scope of the meeting. Specialized as it was, one fact which continued to be repeated during the meeting was one of the common problems of interdisciplinary work; namely, the variability of terminology even between disciplines engaged in the same research area. Although very basic, such a problem was startling and is one which ought to be overcome in the near future.

We would like to stress that the workshop was only possible through financial support from the Stiftung Volkswagenwerk and the cooperation of Zeiss. Further support was provided by the Universitätsbund Göttingen, the Stadtsparkasse Göttingen, and Wild Leitz Inc.

We also owe our thanks to those who contributed at the workshop, and to those who put pen to paper for this volume. Dipl.-Biol. Petra Zimmermann, München, helped us during the final stages of file processing and layout.

Gisela Grupe and Neil Garland

München and Manchester, January 1992

Contents

Invited Participants – Referees *

DR. PAOLO BIANCO *, Dipartimento di Biopathologia Umana, Sezione di Anatomia Patologica, Università "La Sapienza", Policlinico Umberto I, Viale Regina Elena 324, I-00161 Roma, Italy

DR. GEORGES BOIVIN *, Institut National de la Santé et de la Recherche Médicale, U 234, Pathologie des Tissues Calcifies, Faculté Alexis Carrel, Rue Guillaume Paradin, F-69372 Lyon Cedex 08, France

DR. BRENDAN F. BOYCE *, Dept. of Pathology, University of Texas, Health Science Center at San Antonio, 7703 Floyd Curl Drive, San Antonio, Texas 78284-7750, USA

DR. HORST BRUCHHAUS, Institut für Anthropologie und Humangenetik, Friedrich-Schiller-Universität, Kollegiengasse 10, O-6990 Jena, Germany

ANGELA CHILD, School of Chemistry and Applied Chemistry, University of Wales, College of Cardiff, PO Box 912, Cardiff CF1 3TB, UK

DR. A. NEIL GARLAND *, Department of Pathology, University of Manchester, Stopford Building, Oxford Road, Manchester M13 9PT, UK

PROF. DR. GISELA GRUPE *, Institut für Anthropologie und Humangenetik, Ludwig-Maximilians-Universität, Richard-Wagner-Str. 10/I, W-8000 München 2, Germany

PROF. DR. LASZLO HARSÁNYI *, Department of Forensic Medicine, University Medical School of Pécs, Sziget 12, H-7643 Pécs, Hungary

PROF. DR. BERND HERRMANN *, Institut für Anthropologie, Georg-August-Universität, Bürgerstr. 50, W-3400 Göttingen, Germany

PROF. DR. FRIEDRICH W. HEUCK *, Hermann-Kurz-Str. 5, W-7000 Stuttgart, Germany

DIPL.-BIOL. SUSANNE HUMMEL *, Institut für Anthropologie, Georg-August-Universität, Bürgerstr. 50, W-3400 Göttingen, Germany

PD DR. FRANZ PARSCHE, Institut für Anthropologie und Humangenetik, Ludwig-Maximilians-Universität, Richard-Wagner-Str. 10/I, W-8000 München 2, Germany

PROF. DR. ARMAND J. DE RICQLÈS *, Laboratoire d'Anatomie Comparée, Equipe de Recherche: Formations Squelettiques et URA 04 11 37 CNRS, Université de Paris VII, 2 Place Jussieu, F-75251 Paris Cedex 05, France

PD DR. MICHAEL SCHULTZ *, Zentrum Anatomie, Georg-August-Universität, Kreuzbergring 36, W-3400 Göttingen, Germany

DR. HOLGER SCHUTKOWSKI *, Institut für Anthropologie, Georg-August-Universität, Bürgerstr. 50, W-3400 Göttingen, Germany

DR. HILDE UYTTERSCHAUT, Laboratorium voor Anatomie en Embryologie, University of Groningen, Oostersingel 69, NL-9713 EZ Groningen, The Netherlands

DR. JENNIFER WAKELY *, Department of Anatomy, University of Leicester, Medical Sciences Building, PO Box 138, University Road Leicester LE1 9HN, UK

An Introduction to the Histology of Exhumed Mineralized Tissue

A. Neil Garland
Department of Pathology
University of Manchester
Stopford Building
Oxford Road
Manchester M13 9PT
UK

Bone is unique among the tissues of the body. Firstly, during life it is self-repairing and can alter its properties and configuration in response to mechanical demand; and, secondly, because of its biochemical make up and micromorphological structure it can withstand hundreds, if not thousands of years of internment in many different burial environments.

Histology is a technique routinely used in modern day pathological practice but little used in studies of exhumed human and animal bone. In contrast to its uses in palaeontology, microscopy has a poor history as a technique in palaeopathology (see de Ricqlès, this Vol.). Although by the last decade of the 1800s the light microscopic features of bone had been described, efficient polarizing microscopy techniques had been developed, as had techniques for the decalcification and paraffin wax embedding, and handgrinding of human bone, archaeological bone histology remained underdeveloped in comparison to the rather misdirected 19th century anthropological and palaeopathological studies which emphasized skull shape and racial evolution. Histology simply did not fit in with the paradigms of the day.

During the later part of the last century the question of the origin of syphilis began to be debated, a debate which still carries on even today!; likewise the diagnostic criteria enabling anthropologists and pathologists to diagnose syphills in dry bones. Thus, it is perhaps fitting that probably one of the first microscope examinations of exhumed human bone was undertaken by a Dr. Prudden, in 1891, of the Laboratory of the College of Physicians and Surgeons of New York. Two adult tibia exhumed from a prehistoric burial site on the Animas River, Colorado were examined. Although the diagnosis of syphilis was suspected on macroscopic examination, the sections showed a chronic periostitis and osteomyelitis.

In contrast to palaeopathology, which, as a discipline, has an ancient history, palaeohistology is relatively more recent and perhaps palaeohistologists are still at the learning stage of what is possible, and what is not possible, defining procedures and defining more of its applications. As this is the first time ever that animal, human and palaeontological histologists have been brought together, I thought it perhaps fitting to begin the workshop by posing a series of questions pertaining to palaeohistology in the present and also in the future. The questions posed will be paralleled with micrographs of exhumed human bone, and certainly the questions will be just as applicable to animal bone as they are to human material.

What is palaeohistology? The first mention of the term palaeohistology was by Moodie (1926), however the first definition of the term was not provided until 1949 by Graf who stated "...It would seem rather natural to apply the word palaeohistology to the examination of microscopic sections of ancient human beings and the recognizing of tissues and cells in such sections".

What of the uses of palaeohistology? Perhaps the commonest use of palaeohistology today is in the study of bone decomposition, and such studies allow us insight into the interactions which have taken place between the bones and the chemical and physical factors and biological agents within the burial environment. In other words, the study of the post mortem changes in exhumed bone allow us to say something about the taphonomic history of the bones. I should, briefly, like to illustrate this use of histology with the study of some exhumed bones from a charnel deposit at Rothwell Parish Church, Northamptonshire (Fig. 1).

The contents of the crypt have attracted interest since they were first discovered around 1700 A.D. The crypt is only one of two surviving in the British Isles, and is therefore important not only in terms of its structure, but also of its contents. My brief, as a palaeohistologist, was with the use of histology to suggest the nature and extent of any bone damage (for full details of the multidisciplinary investigations see Garland et al., 1988). All the histological sections obtained revealed, to a varying extent, some degree of preservation of the microstructure. However, many aberrations from such a structure were found. In the outer third of the cortex of many of the specimens, generalised destructive changes were found, characterized by a general loss of micromorphology, predominantly affecting the outermost aspect of the osteons (Fig. 2). Focal destructive changes were characterized by focal loss of mineral and collagen and with zones of hypermineralization surrounding the foci (Fig. 3). Loss of organic matrix was seen in the decalcified sections as broken up Haversian systems. Inclusions of fungi were present, lying amongst the trabeculae and within the Haversian systems (Fig. 4). Infiltrations of iron were found in the outer third of the cortex, as demonstrated by a positive Prussian blue reaction (Fig. 5).

The histological study has demonstrated a number of points in the taphonomic history of this collection of bones. There has been destruction and alteration of both the organic and inorganic components of the bones; slow post mortem demineralization is taking place with some of the mineral being reprecipitated inside the bones; loss of organic and inorganic components has led to shrinking and cracking of the bone microstructure; intrusive fungi are present inside the bones and they are continuing and accelerating the destructive processes. Histology has demonstrated a number of important implications for the conservation of both the bones and of the crypt (Garland 1987, 1989).

Virtually all the samples used for bone decomposition studies are taken from the diaphyseal region of the long bones. What of the articular ends? We tend, perhaps, to forget that a joint is a functional unit comprised not only of contiguous ends of bones but also of a cartilaginous component. Does the cartilage survive burial? What are the implications for the preservation of such a layer for bone taphonomy? Is it a barrier against fungal and bacterial invasion? Does the histological preservation of calcified cartilage together with the subchondral bone have a role in the palaeopathological diagnosis of joint disease?

In addition, it must not be forgotten that in general people bury bodies, they do not bury bones. During the early stages of decomposition, and particularly if the body is interred in a closed coffin, the body will be covered with the degradation products of the soft tissue. Do these degradation products together with the changing bacteria and fungi affect bone? Do they in any way initiate bone decomposition? These early changes clearly require investigation which has not yet been adequately carried out (Garland and Janaway 1989).

The next use of palaeohistology to be mentioned is its role as an aid in the diagnosis of disease in exhumed bones. During a symposium on human palaeopathology, held in 1965, two noted osteoarticular pathologists were cynical with regard to the use of histology. Putschar (1966) stated "..one should not, however, expect too much help from the relic, since diagnostic bone patterns are rare. The frail trabeculations of woven bone are probably not sturdy enough to survive long burial". Johnson stated "..The pumice of Paget's disease will not remain, and much of the mosaic will have disappeared". Such dogmatic statements certainly obviscate the field and should be treated with a healthy degree of scepticism. Does woven bone and pagetic bone survive burial?

Ground sections were made of a fragment of a adult male mediaeval skull, excavated in London, suspected on macroscopical grounds to have Paget's disease. The bone, histologically, is well preserved and allowed a diagnosis of Paget's disease to be made (Fig. 6). On the other hand, ground sections and microradiographs made from another skull, again suspected of showing Paget's disease, from a Romano-British site, showed the reverse; there was virtually complete obliteration of the microstructure. However, the decalcified sections suggested some increased scalloping of the cement lines and a slight mosaic pattern.

Radiographs of a humerus excavated from a Saxon site showed the features of monostotic fibrous dysplasia (Wells 1963). Ground sections and microradiography revealed the cortical bone to be virtually filled with focal destructive lesions, although there were areas of the external circumferential lamellae which were unaffected (Fig. 7a). However, in the subcortical bone there were irregular spicules of woven bone still preserved (Fig. 7b). Woven bone has also been identified in histological sections of an osteoid osteoma (Saxon period), a probable osteogenic sarcoma (pre-1778 A.D.), (Fig. 8) and also from fracture calluses (Fig. 9).

Obviously, we can only establish whether woven bone and pagetic bone survive burial by microscopy of the specimen. This leads to several other questions which need to be answered before palaeohistology becomes an accepted technique, particularly in Great Britain. Who decides whether a bone should be sectioned? Where in the sequence of the investigations of a palaeopathological specimen should histology be undertaken? Perhaps answering one type of question with histology may destroy the possibility of answering another with a different technique.

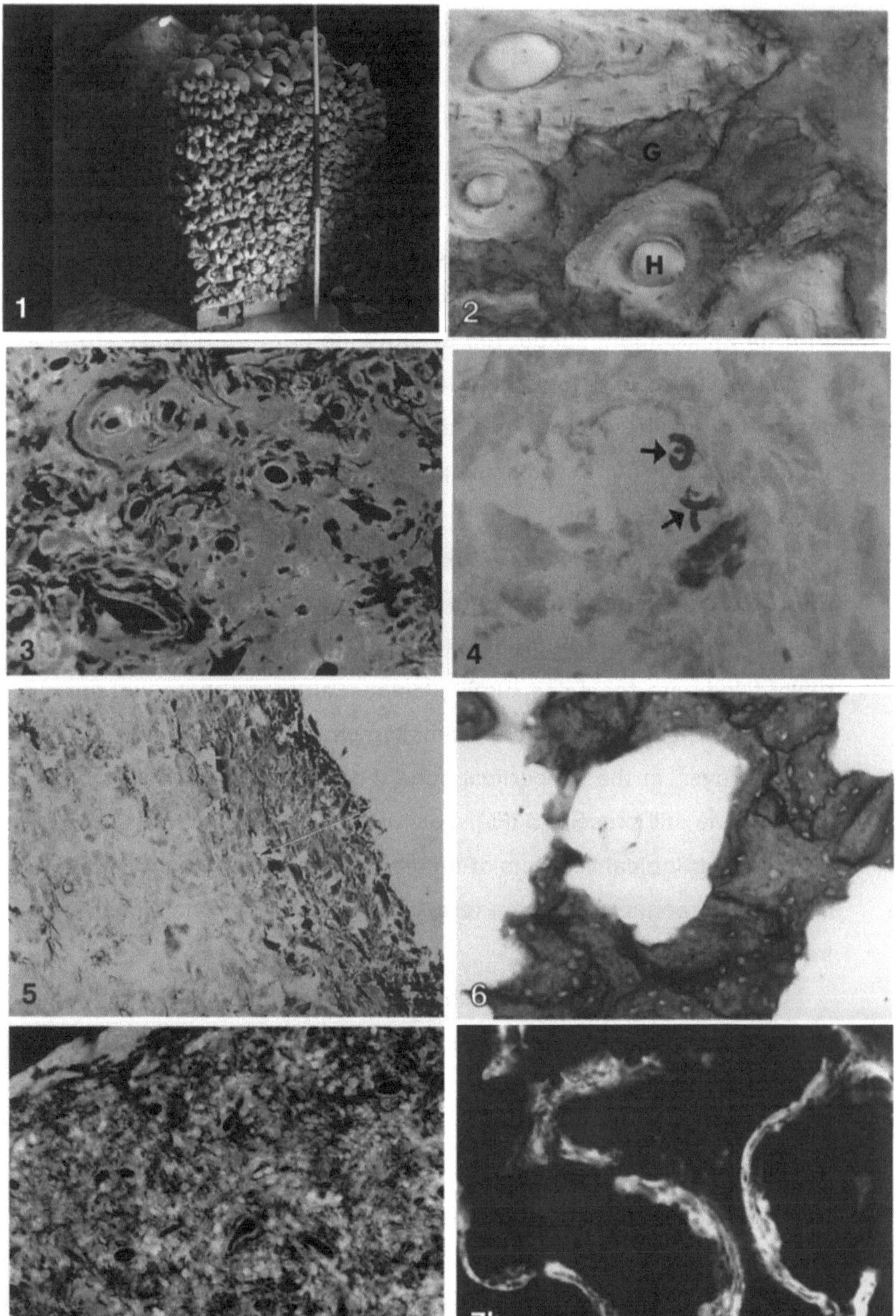
1
2
G
H
3
4
5
6
7a
7b

Fig. 1 View of one of the bone stacks from the charnel house at Rothwell, Northamptonshire. Photo courtesy of Dr. R.C.Janaway, University of Bradford

Fig. 2 Undecalcified unstained section from an adult femur. Generalised destructive changes (**G**), Haversian Canal (**H**); original magnification x50

Fig. 3 Focal destructive changes in the cortical bone of an adult femur; microradiograph, original magnification x30

Fig. 4 Fungal inclusions (arrows) in trabecular bone; PAS stain; original magnification x30

Fig. 5 Infiltrations of iron in the outermost layer of cortical bone (arrow). Prussian blue reaction; original magnification x30

Fig. 6 Undecalcified section of skull showing "scalloped" cement lines. H&E stain; original magnification x25

Fig. 7 a Virtually complete destruction of cortical bone in a humerus with monostotic fibrous dysplasia. Microradiograph; original magnification x25.

Fig. 7 b Irregular spicules of woven bone in the subcortical bone.

Everyone who works on exhumed human and animal bone knows the difficulty in distinguishing ante- from post-mortem changes. Can we confidently diagnose ante-mortem pathology from a purely macroscopic inspection of the specimen when what we are really seeing is the result of so-called pseudopathology? Can histology aid in such differentiations?

Figure 10a shows the endocranial surface of the occipital bone of a child aged 8-10 years at death, excavated from Lasswade, Scotland and dating to 16th century A.D. The bones overlying the superior sagittal sinus, the right and left transverse sinuses and the right and left cerebellar fossae have a "crazy paving" or coralline appearance. However, prior to cleaning those areas of bone were densely covered with fine plant rootlets, and it was thought, perhaps, that the appearance of the bone was the result of etching by plant rootlets. How wrong the archaeologist was. The histological sections (Figs. 10b,c) showed outgrowths of extra cortical new bone formation and good preservation of woven bone as a result of ante-mortem pathology and not post-mortem damage, as was initially thought.

The occipital bone of a skull from Rushen Abbey, The Isle of Man (12th-13th century A.D.) was submitted for investigation. The inner surface of the skull showed a triangular-shaped area of flaking of the surface, which surrounded an area with a honeycombed appearance. Examination of the slab and the slab X-ray showed a virtual loss of the inner table of the skull. The skull was submitted for histological examination to assess the nature of the changes to the inner table of the skull. Were they ante-mortem or post-mortem changes?

Histologically, four zones were evident. A few focal destructive lesions were seen in the external circumferential lamellae (zone 1), but the micromorphology was mainly preserved. Although Haversian canals were identifiable, the remainder of the bone of the outer table (zone 2) was virtually full of focal destructive lesions. The diploe (zone 3) was less densely tunnelled than zone 2 and some micromorphology was recognizable. Soil material was present lying amongst zones 2 and 3. Zone four was represented by a thin strip of cortical bone of the inner table only. In this zone the changes were mainly of the generalized destructive type, but a few tunnels were present in the more densely mineralized areas. This time histology has demonstrated that the changes were post-mortem.

Does histology have a role in the analysis and identification of exhumation associated masses? Numerous "mineralized" masses have been recovered from archaeological excavations worldwide (e.g. Boross and Nemeskeri 1963; Beck and Mulvaney 1966; Baud 1972; Houghton 1975; Strouhal and Jungworth 1977; Strietz et al. 1981; Kramer et al. 1983; Kramer 1984), and they generally take one of two forms. Either they are of a regular shape which often resembles a tissue structure, or they are of an irregular amorphous mass. The four specimens I wish to discuss were excavated from sites containing human remains deposited as a result of two different funerary rituals: (1) inhumation and (2) cremation followed by placing in an urn.

Specimen 1: The mass was recovered from the neck region of a female, age 25-35 years, from the Romano-British site at Poundbury, Dorset. Other pathology was described as "OA of pubic symphysis, thoracic vertebrae and shoulder". The specimen was an irregularly shaped and slightly porous piece of pale yellow coloured material, measuring 19 x 5 mm. The specimen was brittle on cutting. Radiography revealed that the tissue was mineralized. Histological examination (Fig. 11) revealed cortical lamellar bone covered in places by well preserved cartilage, which, however, did not extend deep into the bone. Tide marks were present, suggesting the specimen to have come from a site of endochondral ossification. Although Haversian systems and cement lines were recognizable, much of the tissue matrix was destroyed by focal destructive lesions, or tunnels. The histological appearances suggested the mass to be an exostosis or an osteophyte from an articular surface which had become detached from its original site of origin.

Specimen 2: Some 20 to 30 cylindrical fragments were recovered from the lower abdomen, pelvis and lower limb of an adult male during excavations at the mediaeval site of Barton-on-Humber. The skeleton was found with the knees flexed, not due to a burial ritual, but probably caused by soft tissue pathology associated with the joint changes of a proliferative erosive enthesopathy (Rogers 1989). The specimen submitted for examination was of a cylindrical shape with a dull yellow colour, diameter 7.64 ± 0.02 mm. Staining with H&E, toluidine blue and PAS was uninformative, however sections stained with elastic van Gieson revealed good preservation of the elastic fibres and duplication of the internal elastic lamina (Fig. 12).

With these two specimens, histology has been of use in the identification of their tissue of origin. However, such an identification with the following two specimens, in the opinion of this author, was not possible.

Specimens 3 and 4: The following two specimens were recovered from the same archaeological site. During the anthropological examination of approximately 2330 cremations recovered from the Saxon site at Spong Hill, Norfolk, three fragments from three different cremation urns were discovered which did not have the appearance of cremated bone. The skeletal details mentioned briefly have been extracted from the unpublished report. The fragment not studied here was examined with XRF, with the following results:

Al oxide:	2.600%
Si oxide:	5.957%
P oxide:	45.868%
Ca oxide:	45.254%
Fe oxide:	0.254%

Both specimens were oval to rounded in shape, with a diameter of approximately 10 mm. The predominant colour was dull yellow orange and the internal colour was grey. The weight of the specimens were 0.73 g and 0.32 g respectively. Both specimens showed cracks radiating from the periphery into the core. Plain radiography showed the specimens to be densely mineralized, more so in the core than at the periphery, but without any apparent structure. Histological processing and sectioning proved difficult. Decalcification of a small fragment resulted in its complete dissolution. Although thick sections were readily cut from the resin-embedded blocks, hand grinding was difficult because bits of the section would break off. This was, presumably, because of the inability to fully impregnate the specimens with resin. Histologically, the specimens had a finely granular and nodular appearance. Examination under polarized light revealed birefringence and inclusions of soil material. Higher power of certain areas showed more finely crystalline or granular material with a pale periphery and a darker core which were in a cobble-stone configuration. In addition to the stains described previously, both these specimens were stained with alizarin red, PAS and the Prussian blue in an attempt to identify their components. The PAS stain showed that the peripheries of the specimens were penetrated by actinomycetes and yeasts. Staining with alizarin red, for calcium, was positive, the stain showing no areas of imperfect penetration. Staining with von Kossa revealed patchy areas of positive staining. Microradiography showed a rather irregular distribution of mineral. X-ray diffraction revealed the sample to be highly crystalline hydroxyapatite, with the formula $Ca_5(PO_4)_3OH$. The mole fraction of calcium in the sample, as estimated by atomic absorption spectroscopy, was found to be 0.30. This value is slightly higher than that calculated from the chemical formula (0.28). However, this slight discrepancy was not surprising, given the variable composition of apatites from biological systems.

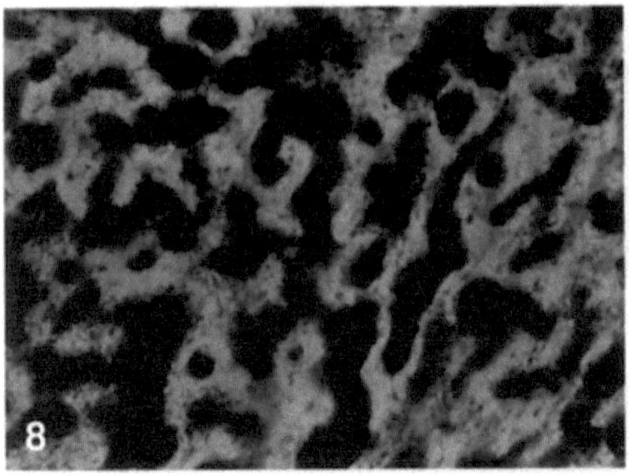
8

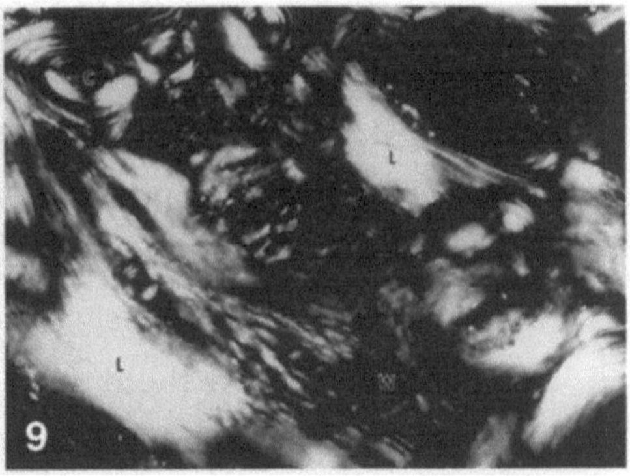
L
L
9

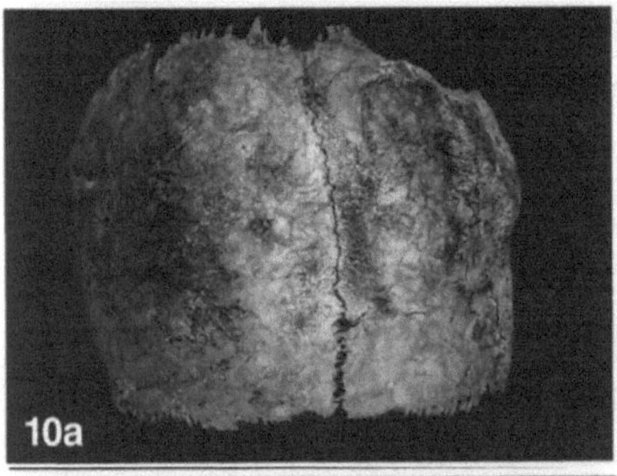
10a

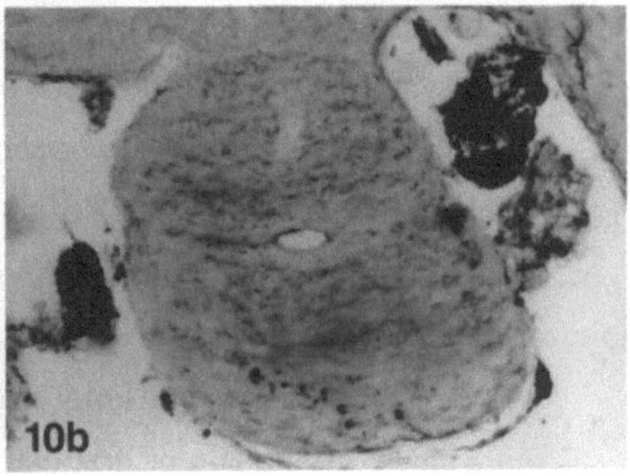
10b

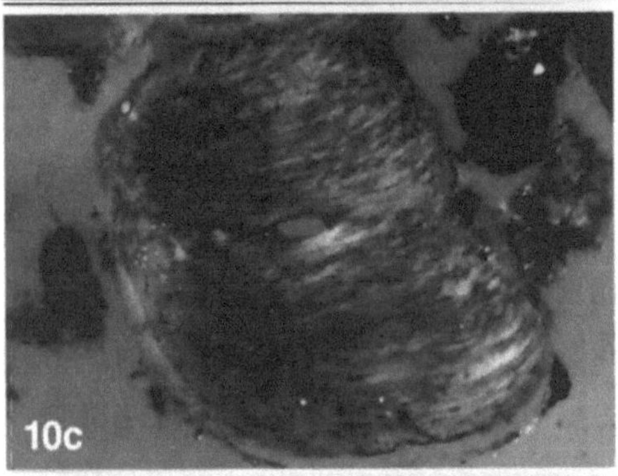
10c

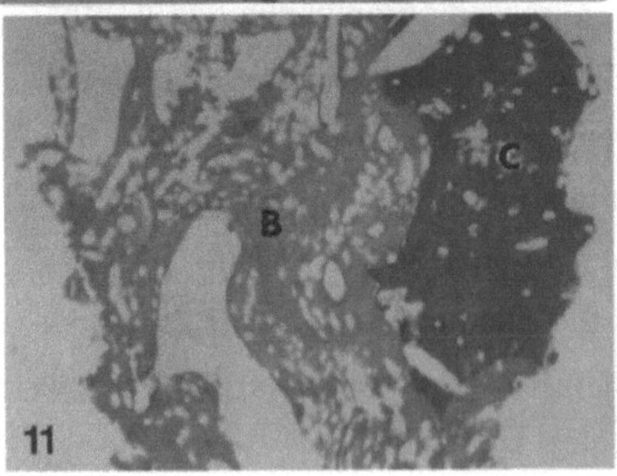
C
B
11

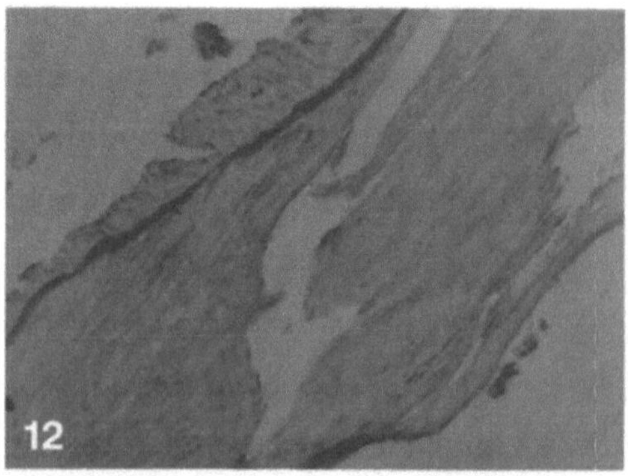
12

Fig. 8 Microradiograph of a specimen of bone from a possible osteogenic sarcoma showing randomly orientated arcades of woven bone; original magnification x25

Fig. 9 Ground section from fracture callus. Woven bone (**W**), lamellar bone (**L**). Polarized light; original magnification x25

Fig. 10 a Endocranial surface of the occipital bone of a child aged 8-10 years. Note the coralline appearance to the bone on either side of the saggital suture.

Fig. 10 b Unstained ground section of above specimen. Note extracortical new bone formation and preservation of woven bone; original magnification x50.

Fig. 10 c Same field, viewed by polarized light

Fig. 11 Preservation of cartilage (**C**) overlying cortical lamellar bone (**B**). Decalcified section; stain: toluidine blue; original magnification x25

Fig. 12 Section of a mineralized blood vessel. Good preservation of elastic fibres; stain: elastic van Gieson, original magnification x25

The examination of cremated remains is a daunting task unfavoured by many workers and so it is not surprising that only limited studies have been undertaken regarding the histological examination of cremated material (Herrmann 1977, 1988; Shipman et al. 1984). The only comparable specimens to the two just described were two spherical concretions recovered from the cremated remains of a male, aged 40-50 years, from an early pre-Roman Iron Age (ca. 500-250 B.C.) cemetery, Schleswig Holstein, FRG (Schutkowski et al. 1986). Macroscopic examination showed them to be composed of distinct layers of differing consistencies and examination under polarized light demonstrated a concentric lamellar structure of the microarchitecture. Microradiography, however, revealed a homogeneous distribution of mineral. X-ray diffraction showed the concretions to have a chemical structure characteristic of apatite.

With the two specimens described above, excavated from Spong Hill, only limited conclusions could be drawn from the histological appearances and a differential diagnosis of kidney or bladder stones or calcified lymph nodes (as a result of a granulomatous disease) has to be proposed.

What of palaeohistology in the future? The main aim of the Palaeohistology Workshop was to discuss the research potential and the practical uses of bone histology as applied to exhumed (human and animal) skeletal material. The following contributions are those discussions. Palaeohistology is a unique and exciting tool for the study of extant material, and its total potential is, as yet, untapped.

References

Baud C (1972) Une plaque pleurale calcifée: étude ultrastructurale et crystallographique. Geneva 20:196-199

Beck SW, Mulvaney WP (1966) Apatitic urinary calculi from early American Indians. JAMA 195:1044

Boross MM, Nemeskeri J (1963) Ein bronzezeitlicher Nierenstein aus Ungarn. Homo 14:149

Garland AN (1987) A histological study of archaeological bone decomposition. In: Boddington A, Garland AN, Janaway RC (eds) Death, decay and reconstruction: approaches to archaeology and forensic science. Univ Press, Manchester pp 109-126

Garland AN (1989) Microscopical analysis of fossil bone. Appl Geochem 4:215-229

Garland AN, Janaway RC (1989) The taphonomy of inhumation burials. In: Roberts CA, Lee F & Bintliff J (eds) Burial archaeology: current research, methods and developments. BAR Brit Ser 211:15-37

Garland AN, Janaway RC, Roberts CA (1989) A study of the decay processes of the human skeletal remains from the Holy Trinity, Rothwell Parish Church, Northamptonshire. Oxford J Archaeol 7:235-252

Graf W (1949) Preserved histological structures in Egyptian and ancient Swedish skeletons. Acta Anat 8:236

Houghton P (1975) A renal calculus from proto-historic New Zealand. OSSA 2:11-14

Jarcho S (1966) The development and present condition of human palaeopathology in the United States. In: Jarcho S (ed) Human palaeopathology. Yale Univ Press, New Haven, pp 3-30

Kramer C (1984) Plaques pleurales chez un homme du Moyen Age: étude radiologique, microscopique et cristallographique. In: Proc 5th Eur Meet Paleopathology Association, Sienna, pp 199-204

Kramer C, Baud CA, Lagier R (1983) Presumed calcified leiomyoma of the uterus. Arch Pathol Lab Med 107:91-93

Moodie RL (1926) Studies in palaeopathology, X111: the element of the Haversian system in normal and pathological structures among fossil vertebrates. Biol Generalis 2:63-95

Putschar WGJ (1966) Problems in the pathology and palaeopathology of bone. In: Jarcho S (ed) Human palaeopathology. Yale Univ Press, New Haven, pp 57-65

Rogers J (1989) Case histories. In: The antiquity of the erosive arthropathies. ARC Conf Proc 5:45-7

Schutkowski H, Hummel S, Gegner, S (1986) Case reports on paleopathology 8. Paleopathol Assoc Newslett 55:11-12

Strietz JM, Aufderheide AC, El-Najjar, MY (1981) A 1500-year old bladder stone. J Urol 126:452-453.

Strouhal E, Jungwirth J (1977) Ein verkalktes Myoma uteri aus der späten Römerzeit in Ägyptisch-Nubien. Mitt Anthropol Ges Wien 107:215-221

Wells C (1963) Polyostotic fibrous dysplasia in a seventh-century Anglo-Saxon. Brit J Radiol 36:925-6

Wells C (1965) A pathological Anglo-Saxon femur. Brit J Radiol 38:393-4

Light Microscopy of Excavated Human Bone

Bernd Herrmann
Institut für Anthropologie
Georg-August-Universität
Bürgerstr. 50
W-3400 Göttingen
Germany

Today's prehistoric anthropology could not produce satisfying results without microscope assistance, either light microscopy or electron microscopy.

Because of the technical difficulties in handling delicate bony substance (Robinson et al. 1987), the impact of electron microscopy on structure research in anthropology has not yet reached its potential. However, light microscopy, at least the present time, is more important not only because of its availability and access to researchers, but because it is simpler.

Simpler? Despite the appearance of a simple opto-mechanical tool, light microscopy, used on a sophisticated scale, requires adequate physical knowledge. This underestimation often leads to poorly selected light microscopy methods, and as a consequence, to insufficient results. On the other hand, the introduction of improved embedding techniques, e.g. by using modern plastic resins and cutting techniques as with the rotation sawing microtome, enabled scientists to concentrate on non-fossilized bones. These techniques have been used on a large scale, only within the last 15-20 years. Of course, important papers have been presented earlier on the microstructure of non fossilized prehistoric human bones by well-known scholars, for example A. Ascenzi or Ch. Baud and others. But it was not until 1966 that the first systematic evaluation of the use of bone histology in forensic medicine and anthropology was presented by D. Enlow. This is considered a benchmark and breakthrough, at least for prehistoric anthropology, since the scientific community became aware of the importance of using this type of information.

However, there is one other paper which is often overlooked that seems to be of similar importance, since it deals with decalcification and staining of archaeological bones and with the histochemical interpretation of metachromasia (Andersen and Jorgensen 1960). Although it did not contribute much to our knowledge, it presented an innovative approach. The authors were able to prove the presence of acid mucopolysaccharides by histochemical methods in old Eskimo and bog bones. Together with an earlier paper by Thunberg, who detected citric acid in old bones already in 1947, this would appear to have been the beginning of (bio)chemical/molecular archaeology (or biomolecular palaeontology), which has developed into an explosively expanding area in bone microtechniques today.

Light microscopy is a tool to obtain information on at least four major branches of prehistoric anthropology (see Fig. 1):

1. It is used to discriminate and describe species characteristics (e.g. human vs non human);
2. It is used to determine decomposition processes (e.g microbial deterioration vs physiological structure);
3. It is used to age the specimen, since biological age is not only fundamental biological information, it is also important with regard to social history features in human societies;
4. It is used for palaeopathological diagnosis, at least when differential diagnoses have to be determined and "pseudopathologies" are to be eliminated.

It is possible to obtain this basic information even with standard microscope equipment. Of course, the results also depend on the methods of preparation (i.e. embedding and cutting), which are not considered here. Being basically an optical task, the histological examination of ancient bone is considered in terms of light microscopy. Light microscopy methods represent an adequate approach to examining the bony substance at the tissue level. The study of smaller tissue components, of either mineral or organic origin, would be more successful using other methods and different magnification levels than those available to standard light microscopy. Thus, the following discussion is restricted to tissue inspection and ignores for the time being organic residues or remains, regardless of whether these are physiological or have been invaded by living organisms. However, the following methods are also useful for the examination of these remains if they are related to tissue structures.

Basically, there is not much difference between the optical impression of a native, old sample and a thin section of a bone from the dissection room, despite the fact that there is a "prestaining" of ancient bones due to soil conditions.

Recommended methods for observation are bright field, dark field, phase contrast, Nomarski differential interference contrast, polarized light and fluorescence light (see note at the end of this paper).

Bright field observation, which is the standard technique, is possible without adding any further attachments. All other microscopy methods are performed by attaching additional modules to the microscope. Dark field observation is possible by inserting a dark field condenser. Now light does not pass the sample directly to meet the observer's eye, but brightness is projected to all parts of the sample where phase differences between light waves appear. A phase contrast attachment is especially useful for minute structural details of objects such as incremental lines. Nomarski differential interference contrast equipment (NDIC) indicates both brightness/darkness differences and interference colours. This certainly produces very sophisticated images but they are not easy to interpret, since edges of the same structure appear in different colours if they oppose each other, moreover, the image contrast depends on the orientation of the sample. However, NDIC is especially useful for relief like stereo images of poorly structured samples on an improved level compared to phase contrast images.

Next to bright field inspection, polarized light is the most common observation method to examine old bones. Adding an analyser and a polarizer to the microscope and orientating polarization planes to a right-angled position leads to bright and dark areas. The dark areas are due to light absorbance by the analyser, while the bright areas are due to anisotropic characters which are within 0° and 90° of the oscillation plane of the light. Adding a ¼ lambda + tint plate to produce interference colours within red first order, one obtains, for example, blue if the directions of parts having higher refraction indices are parallel to that of the tint plate, whereas yellow indicates directions of higher refraction indices at a 90° angle to the tint plate. Thus, if the sample, the plate or the polarizer is rotated, the colours will change and basic information about the physical properties of the sample inspected is obtained.

As research is presently moving into the field of molecular compounds in ancient bones transmitted and reflected light fluorescence attachments will also certainly provide very useful information. Unfortunately only few compounds have the natural ability to fluoresce (Piepenbrink et al. 1983). Fluorescence microscopy will become increasingly important because of its high specificity and high sensitivity in proving the presence of compounds in old bones after treatment with specific dyes and biomolecular probes.

If all the above mentioned methods are applied to prehistoric bones, the results mostly depend on their state of preservation. In case of reasonable preservation the differences in appearance compared to a modern sample are small. Differences and strange impressions increase with bone decomposition.

As mentioned above regarding the fluorescence technique, histological inspection of old bones makes one aware of its usefulness for molecular approaches. Assessment of the state of deterioration of a bone certainly prevents wasting time on an unsuitable sample. Another field of application still awaits exploration. It seems apparently that at present not much systematic work has been carried out regarding the use of dyes on prehistoric bone.

A few tests were done in advance of this workshop to demonstrate how promising further work in this area would be. The staining methods were chosen with respect to their availability to any histological lab. Tests were done with Masson-Goldner, Azan, KPM, haematoxylin and eosin, Romanowski-Giemsa, Villanueva, methylene blue and crystal violet.

This selection obviously includes also allochromasia, metachromasia and trichromic dyes and stains. The staining methods were applied on four samples from prehistoric bones that differed in age and environmental conditions (Table 1). Samples were selected from femora of young adult individuals. Neighbouring sections were stained with different stains, thus revealing their influence on almost identical microstructures.

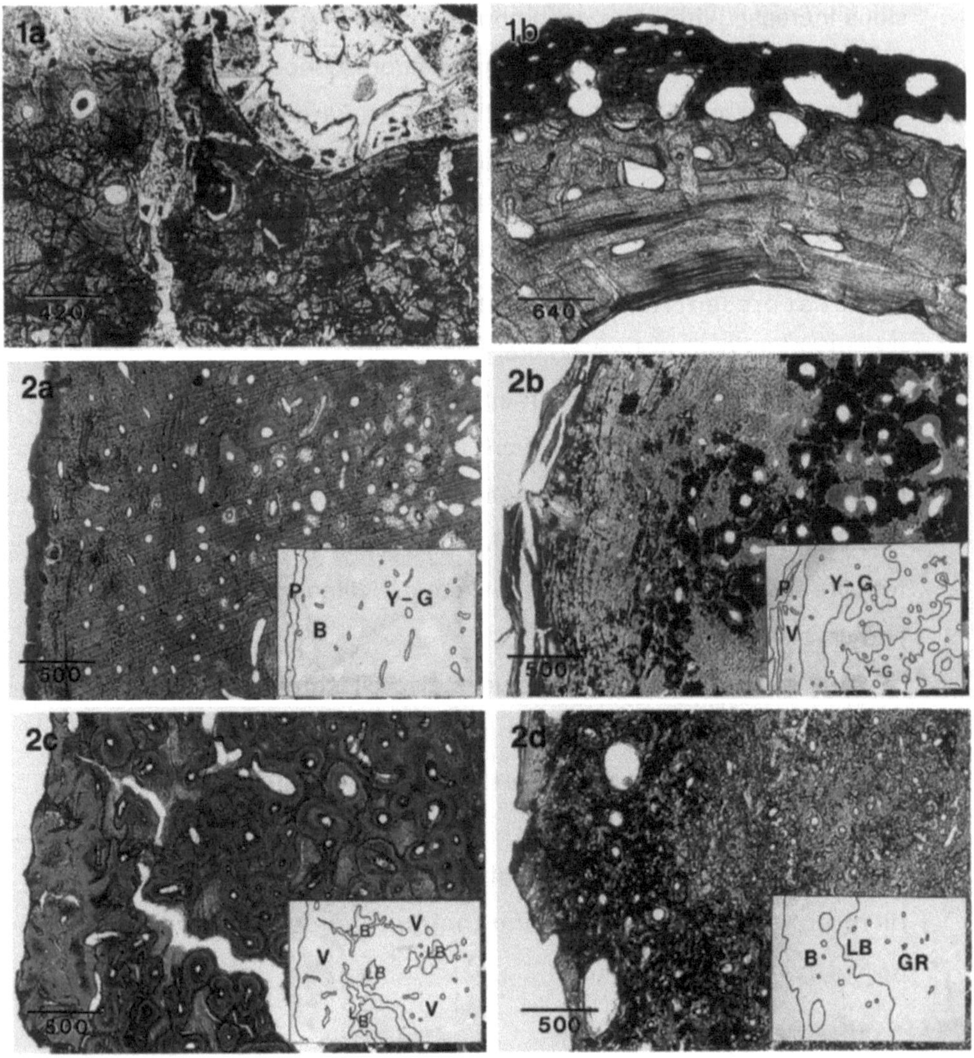
1a
420
1b
640
2a
500
P
Y-G
B
2b
500
P
Y-G
V
Y-G
2c
500
V
LB
V
LB
V
LB
B
V
2d
500
B
LB
GR

Fig.1 a,b The four main features of information from histological examination of prehistoric bones. As shown in **a**, determination of the species characteristics and state of preservation might be assessed. Cross section through a bone most probably resembling a humerus from Swartkrans Cave deposits, tentatively addressed as *Australopithecus* sp. Diagenic changes in osteons and conversion of bony tissue into mineral phases (**above**) is detectable; bright field.
In **b**, ageing and pathological features might be judged. Cross section through a femur of an Infans I individual from the Espenfeld site (mediaeval; Thüringen, FRG) with pathological apposition of bony material due to periostal reactions of unknown aetiology; bright field with less polarized light; magnification bar in μm

Fig. 2 Cross-sections of the femora from young adult individuals of different age and different states of preservation due to different soil conditions. All samples stained simultaneously as described in the text with KPM stain (Dane and Herman 1963). Sketches are for colour patterns. Magnification bar in μm.

Fig. 2 a Sandy soil; mediaeval (City of Münster, FRG). Periostal parts purple (**P**), **B** brownish, turning to yellowish-green.

Fig. 2 b Sand/clay/dung; mediaeval (City of Schleswig, FRG). Periostal area purple (**P**) and vermillion (**V**), followed by yellowish-green (**Y-G**); osteons appear dark green to black, enclosing areas of Y-G.

Fig. 2 c Regosol, 3rd millennium b.c. (United Arab Emirates). Section stained homogeneously vermillion (**V**), interrupted by only a few light blue areas (**LB**). **d** Loess; 3rd millennium BC (Thüringen, FRG). Periostal part brown (**B**), other part light blue (**LB**) and grey (**GR**)

Table 1 Details of specimens

Samples

Modern	Dissection room		
Mediaeval	Sandy soil	12-15 Century a.d.	Münster, FRG
Mediaeval	Sand/clay/dung	11-13 Century a.d.	Schleswig, FRG
Late chalco-lithic/ Bronze age	Regosol	3 million b.c.	United Arab Emirates
Neolithic	Loess	3 million b.c.	Thüringen, FRG

The tests were carried out as follows:

1. Optimal processing for each staining method was determined by pretests.
2. Thin sections of all samples were stained at the same time in the same staining solution for the same length of time. It was assumed that any difference in appearance of stained sections by a given staining method should be caused by either chemical or structural decomposition, or both.
3. The results varied broadly, but why this is so is not yet understood. Of course, an explanation is even more difficult since the behaviour of the dyes themselves is not well understood. The most remarkable differences were found by applying the KPM method to old samples. Although it was originally introduced as a method for keratin-prekeratin-mucin staining in skin sections (Dane and Herman 1963), it proved to be an excellent stain for old bones (Fig. 2).

From these tests only the results from the KPM set are presented here, since they were basically the same in all tests. Comparing the samples in Fig. 2 it is obvious that environmental conditions are most important. There seems to be an advantage in using staining methods to screen old bones with respect to their biochemical and molecular components. These general staining methods as applied here are probably not very advantageous because of their low histochemical specificity. This could easily be enhanced by using specific histochemical agents (Kiernan 1990). One would also expect the introduction of antibodies as probes to improve the specificity. In this field and in the application of morphometric methods (e.g. Frost 1987) to the histology of old bones, we can expect dynamic development within the next few years. Together both certainly add a further major topic to the four mentioned above in future work.

Note

This paper presented to the workshop was illustrated by numerous colour slides. Including them here as black and white images would not contribute equally. Technical assistance by Jens Kerl and Ursula Schulz is gratefully acknowledged. Preparation of slides presented at the workshop was made possible by the courtesy of Mr. Junker and Mr. Ernst of Olympus Optical Co., Hamburg (FRG).

References

Andersen H, Jorgensen J (1960) Decalcification and staining of archaeological bones, with histochemical interpretation of metachromasia. Stain Technol 35:91-96

Dane E, Herman D (1963) Haematoxilin-Phloxin-Alcianblue-OrangeG. Differential staining of prekeratin, keratin and mucin. Stain Technol 38:97-101

Enlow D (1966) An evaluation of the use of bone histology in forensic medicine and anthropology. In: Evans FG (ed) Studies on the anatomy and function of bone and joints. Springer, Berlin Heidelberg New York, pp 93-112

Frost HM (1987) Secondary osteon populations: an algorithm for determining mean bone tissue age. Yearb Phys Anthropol 30:221-238

Kiernan JA (1990) Histological and histochemical methods: theory and practise, 2nd edn. Pergamon, Oxford

Piepenbrink H, Herrmann B, Hoffmann P (1983) Tetracyclin-typische Fluoreszenzen in bodengelagerten Skeletteilen. Z Rechtsmed 91: 71-74

Robinson D, Ehlers U, Herken R, Herrmann B, Mayer F, Schürmann F W (1987) Methods of preparation for electron microscopy. Springer, Berlin, Heidelberg, New York

Thunberg T (1947) The citric acid content of older, especially mediaeval and prehistoric bone material. Acta Phys Scand 14: 245

Decomposition Phenomena in Thin Sections of Excavated Human Bones

Gisela Grupe and Ute Dreses-Werringloer
Institut für Anthropologie und Humangenetik
Ludwig-Maximilians-Universität
Richard-Wagner-Str. 10/I
W-8000 München 2
Germany

1 Introduction

When the soft tissue of a corpse has gone, the bone is directly exposed to its burial environment. Although mineralized tissues are capable of surviving hundreds or thousands of years in the soil, not all specimens are likely to become fossils. Most of them will be subsequently decomposed until nothing remains but a "soil silhouette". Therefore, every archaeological bone has experienced some destruction to a major or minor degree. It is most important to differentiate decomposition phenomena from physiological features of the specimen with certainty, otherwise any diagnosis of species, individual age or pathology might be biased or simply false. For example, even well-preserved bones frequently reveal superficial destruction (Fig. 1) which might be mistaken for an eroded hyperostosis. Or, as another example, a poorly mineralized specimen may be due to a metabolic disease or simply the result of mineral loss by leaching after burial. The problem of different decomposition processes and velocity in healthy and pathological specimens was recently addressed by Bell (1990). It is also obvious that the agent which produces decomposition phenomena should be known.

There are three principal methods of bone destruction which, however, show considerable overlap and any sharp distinction between them appears a little artificial. The physical decomposition process of bone weathering is a well-known phenomenon and results in easily detectable microfissures and crackings. They are mainly brought about by sediment pressure and temperature, but also, in the case of superficial finds, by trampling. Their main importance lies in the fact that any cracking provides an additional interior bone surface which enhances further decomposition.

Cracking also results when the formation of new calcium phosphate crystals begin after initial hydrolysis of the original bone apatite (Fig. 2). Chemical bone destruction is a more complicated process and is mainly dependent upon soil pH, soil drainage, the local redox potential and other features of the burial environment (e.g. Gordon and Buikstra 1981, Williams 1988). Minerals may be leached from the specimen, others may contaminate it and become fixed by ionic substitution. Initial hydrolysis of the bone mineral is frequently followed by recrystallization which also involves the inclusion of extraneous material into the bone (e.g. Pate et al. 1989).

The process of biological decomposition is, by far, less understood. It is mainly due to microbial activity that the collagen and non-collagenous proteins of the bone are destroyed by several species of soil fungi and soil bacteria. In the case of submerged finds, blue-green algae should be considered as being of importance. Any degradation of the organic matrix of a bone at the same time destroys the framework for the mineral crystals (White and Hannus 1983, Von Endt and Ortner 1984) and thus also initiates the decomposition of the mineral matrix. Focal destructive lesions (tunnels) in excavated bones are thought to be the result of microbial activity (Fig. 3), whereby the demineralisation of the bone is enhanced by acidic microbial metabolites (e.g. Marchiafava et al. 1974; Piepenbrink 1989). In fact, some of the soil microorganisms which are presumably responsible for biological bone decomposition (see below) are capable of solubilizing phosphate minerals (e.g. Sperber 1958; Duff et al. 1963; Metha and Bhide 1970). At this point, the bone again enters the chain of chemical decomposition processes, i.e. leaching, hydrolysis, recrystallization and so on.

Besides obvious signs of weathering, post mortem impacts on a bone's integrity are indicated by stains and, histologically, by a loss of birefringence and the presence of tunnels. Stains can be the product of mineral contamination (e.g. blue colour after vivianite formation), but other colours cannot be correlated with extraneous metal incorporation. Frequently, excavated bones show areas coloured red, blue or violet, whereby the stains tend to oxidize when exposed to air and turn the bones black. Since the coloured areas are accompanied by erosion and cracking (Piepenbrink 1984), the phenomenon might be mistaken for heat exposure. Those stains are assumed to derive from microbial metabolites, but there is presently almost no clear relationship between the action of microorganisms and the alteration of bone features (with a few exceptions, e.g. Marchiafava et al. 1974).

It is also assumed that tunnelling of bone is caused by soil microorganisms, presumably by soil fungi. Experimental work has shown that a great deal of decompositional change is, at least, initiated during the early stages of diagenesis (Grupe and Piepenbrink 1989), and, since bone is only attractive to microorganisms when a substantial amount of organic material is still left, a current research focus lies on microbial dead bone decomposition. The approach to the subject is both phenomenological and experimental.

2 Material and Methods

Anthropological thin sections taken from the shaft of a human long limb bone were investigated by light microscopy for the purpose of age diagnosis; thin sections from modern pig bones which had been experimentally inoculated with soil microorganisms were also examined. In addition, a fluorescence light microscope with UV-excitation and a light yellow 435 nm barrier filter was used.

For thin sectioning, compact bone was embedded into the resin "Biodur E12" and sectioned with a standard horizontal rotation sawing microtome (Leitz 1600) to a thickness of approximately 80 μm. For the detection of microbial remains in the bones, the embedded sections were stained by non-specific cell stains. They were stained either with methylene blue for 5-15 min, or with ammonium oxalate/crystal violet for 1-5 min.

The pig bone specimens for the inoculation experiments were macerated in cold water and then sterilized by uranium radiation (dose: 25 kGy). The following microorganisms were found to be capable of living on sterilized bone from the organic remains:

Soil fungi:	Soil bacteria:
Botrytis sp.	
Cladosporium cladosporioides	*Actinomadura madurae*
Cladosporium herbarum	*Alkaligenes pichaudii*
Cladosporium sp.	*Arthrobacter globiformis*
Crysosporium pannorum	*Bacillus subtilis*
Crysosporium sp.	*Pseudomonas fluorescens*
Penicillium brevi-compactum	*Streptomyces griseus*
Penicillium herbarum	
Phialophora hoffmannii	
Scytalidium sp.	

The soil fungi had been extracted from soil adjacent to buried bones (Piepenbrink 1984), the choice of bacteria was due to their ability to cleave proteins and to produce pigments. No experiments were carried out with the hazardous bacterium *Clostridium* because of a lack of the necessary safety equipment. These bacteria, however, are unlikely to participate significantly in bone decomposition due to their very narrow temperature optimum (Angela Child, pers. comm.).

The bones inoculated with fungi were kept in the dark, at 4°C (minimal average grave temperature) and in aerobic conditions. The bacteria were also capable of surviving at lower temperatures, but the inoculated bones were kept at their optimal growing temperatures simply to enhance the reaction. After approximately 4 months the compact pig bone inoculated with fungi or bacteria were densely colonized.

An explanation should be added as to why we tried to detect microbial tissue in thin sections of bone by light microscopy. Bacteria are very small (1-4 μm, depending upon their state of growth and environmental parameters), thus any SEM or TEM technique would have been more suitable. However, as excavated skeletons are routinely thin sectioned by anthropologists for the evaluation of individual biological parameters (for example, age and pathology), as a matter of economy the same sections should be used for the detection of decomposition phenomena, especially if the individuals under study are also to be further investigated biochemically(e.g. for trace element or stable isotope analysis, DNA extraction, or radiocarbon dating).

3. Microbial Remains in Thin Sections

Fungal invasion of bone is easily detected since the organisms grow as hyphae of a substantial size. It is important to note that fungal remains are very difficult to remove from an infected bone (Grupe and Piepenbrink 1989). None of the pig bones inoculated by us with fungi ever revealed tunnelling, but in excavated bones, hyphae-like structures are visible after staining the sections with a cell stain (Fig. 4). It is therefore hypothesized that tunnelling may be a quite late stage of bone diagenesis: the acidic metabolites partly dis-

Fig. 1 Microradiograph of a thin section through the midshaft of a human phalanx; early mediaeval specimen from Altenerding, southern Germany. Note fissures and crackings at the eroded surface as well as a demineralization front (*arrow*). Areas of increased radiodensity represent recrystallization products (inside the Haversian systems and eroded surface areas). Scale in μm

Fig. 2 Thin section through the midshaft of a human femur; mediaeval specimen from Schleswig, northern Germany. As a result of an excess of Fe-ions in the soil, vivianite crystals [$Fe_3(HPO_4)x8H_2O$] are formed. This space consuming process leads to bone cracking. Scale in μm

Fig. 3 Thin section through the midshaft of a human femur; mediaeval specimen from Münster, Germany. The location of the tunnels follow the original circumferential bone layers of the Haversian systems. The cavity of the former blood vessel contains four round, naturally stained structures (*arrow*), presumably preserved parasite remains. Scale in μm

Fig. 4 Compact human bone, stained with methylene blue; mediaeval specimen from Espenfeld, Germany. The thin section shows tunnelled areas next to intact microstructures. The cell stain is both accepted by preserved osteocytes (*areas to the left*) and other condensed material within or next to the tunnels. Hyphae-like structures are also found either within the tunnels or seemingly originating from them. Scale in μm

Fig. 5 Human metatarsal of a mediaeval leprous skeleton from Odense, Denmark, stained with methylene blue. Three of the tunnels are densely filled with condensed material which accepted the dye (*arrows*), the others are void. Scale in μm

Fig. 6 Thin section through compact pig bone inoculated with *Actinomadura madurae* and stained with methylene blue. Fluorescence light image. The stained microbial material is located within a natural micropore. Scale in μm

Fig. 7 Compact pig bone inoculated with *Pseudomonas fluorescens* and stained with crystal violet. Fluorescence light image. Bacterial cells and colonies are visible within natural micropores as well as within an artificial crack (*arrows*). Note the bright fluorescence of microbial metabolites around the infected areas. Scale in μm

Fig. 8 Fluorescent microbial metabolites are invading the bone by diffusion. Unstained pig bone inoculated with *Actinomadura madurae*. Microbial tissue is visible as a thin layer at the rim of the natural pores (*arrows*). Scale in μm

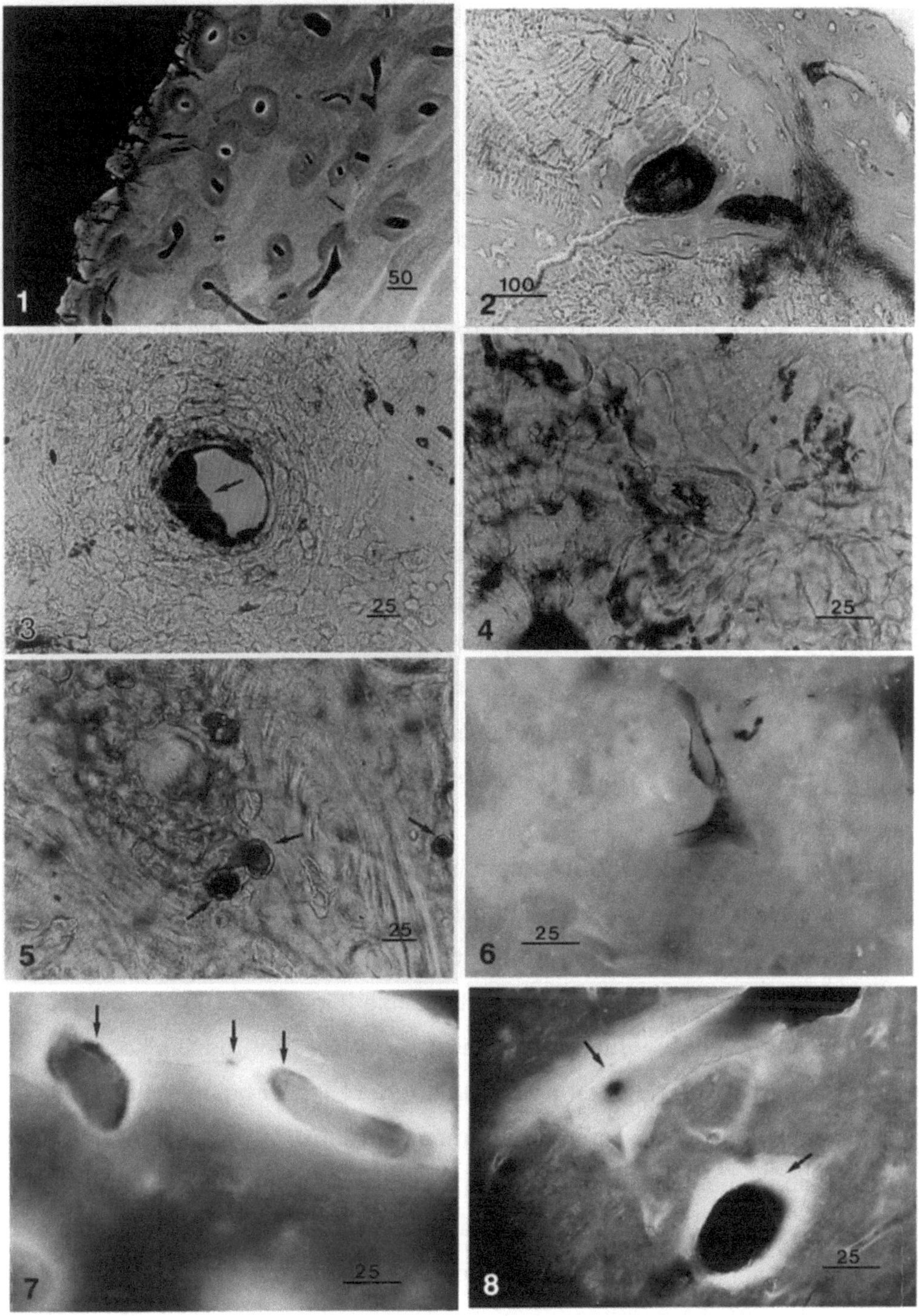
1
50
2
100
3
25
4
25
5
25
6
25
7
25
8
25

solve the mineral matrix, which is later washed away by soil humidity percolating through and around the buried skeleton. The incubation time of some months is probably insufficient to mimic the whole biological decomposition process.

It is impossible to identify the stained exogenous organic remains to the species level. Other sections demonstrated that the stained material is itself greatly altered, appearing as a dense, amorphous mass. Since the invading microorganisms themselves die and decompose as well, it is not surprising that some of the tunnels were even void (Fig. 5).

It was not possible to identify bacterial colonies or even single bacterial cells in excavated bones due to their small size. On the other hand, the inoculation experiments revealed that the organisms were capable of growing on the bones, and they even left their pigments there. Bones infected with *Pseudomonas fluorescens* tended to appear a neon-like yellow after the incubation period. This colour, however, is due to a water-soluble pigment and is unknown for excavated skeletons. The hyphae of *Actinomadura madurae* stained the bone brown within only a few weeks.

The detection of bacteria in inoculated bones was successfully undertaken by the use of a fluorescence light microscope. In so doing, most of the indigenous bone structures are depressed, and the stained microorganisms are clearly visible (Figs. 6 and 7). *Pseudomonas fluorescens*, which was one of the most "aggressive" bacteria used in the experiments, appeared bright red after being stained with crystal violet. This could be an important hint for the explanation of metachromasia effects in ancient bones (see Herrmann, this Vol.): some exogenous organic material is to be expected in every ancient bone specimen, so the acceptance of a dye by the microbial remains can be at least partly responsible for the observed metachromasia effects.

The implications of this study concern not only the possible artefacts in palaeohistological specimens, but are also very important for the dating of ancient bone finds or other bone chemistry studies. After only some months, the experimentally inoculated bones were densely colonized with fungi or bacteria, but did not reveal the slightest loss of their integrity. In addition, this means that ancient specimens which appear to be very well preserved can already contain a huge amount of exogenous material. This material must not necessarily be restricted to natural micropores or tunnels, but can also be transported through the bone by diffusion (Fig. 8). Microbial tissue contains amino acids, lipids and proteins and thus provides possibly sufficient carbon, nitrogen and oxygen to confuse biochemical bone studies.

The search for microbial remains in excavated bones is also likely to add fruitful information in cases of possible tetracycline incorporation by infected grain consumption which is still a matter for discussion among palaeopathologists (e.g. Cook et al. 1989).

Acknowledgements. The experimental work was financially supported by the Deutsche Forschungsgemeinschaft.

References

Bell LS (1990) Palaeopathology and diagenesis: an SEM evaluation of structural changes using backscattered electron imaging. J Arch Sci 17:85-102

Cook M, Molto E, Anderson C (1989) Fluorochrome labelling in Roman period skeletons from Dakleh Oasis, Egypt. Am J Phys Anthropol 80:137-143

Duff RB, Webley DM & Scott RO (1963) Solubilization of minerals and related materials by 2-ketogluconic acid-producing bacteria. Soil Sci 95:105-114

Gordon CC, Buikstra JE (1981) Soil pH, bone preservation, and sampling bias at mortuary sites. Am Antiqu 46:566-571

Grupe G, Piepenbrink H (1989) Impact of microbial activity on trace element concentrations in excavated bones. Appl Geochem 4:293-298

Marchiafava V, Bonucci E, Ascenzi A (1974) Fungal osteoclasia: a model of dead bone resorption. Calc Tissue Res 14: 195-210

Metha YR, Bhide VP (1970) Solubilization of tricalcium phosphate by some soil fungi. Indian J Exp Biol 8:228-229

Pate FD, Hutton JT, Norrish K (1989) Ionic exchange between soil solution and bone: toward a predictive model. Appl Geochem 4:303-316

Piepenbrink H (1984) Beispiele biogener Dekompositionserscheinungen an Knochen unter längerer Liegezeit. Anthropol Anz 42:241-251

Piepenbrink H (1989) Examples of chemical changes during fossilisation. Appl Geochem 4:273-280

Sperber JI (1958) Solution of apatite by soil microorganisms producing organic acids. Aust J Agric Res 9:782-787

Von Endt DW, Ortner DJ (1984) Experimental effects of bone size and temperature on bone diagenesis. J Arch Sci 11:247-253

White EM, Hannus LA (1983) Chemical weathering of bone in archaeological soils. Am Antiqu 48:316-322

Williams CT (1988) Alteration of chemical composition of fossil bones by soil processes and ground water. In Grupe G, Herrmann B (eds) Trace elements in environmental history. Springer Berlin Heidelberg New York, pp 27-40

Some Remarks on Palaeohistology from a Comparative Evolutionary Point of View

Armand J. de Ricqlès
Laboratoire d'Anatomie Comparée
Equipe de Recherche: Formations Squelettiques
et URA 04 11 37 CNRS
Université de Paris VII
2 Place Jussieu
F-75251 Paris Cedex 05
France

> "Full fathom five thow Father lies;
> of his bones are corals made;
> Those are pearls that were his eyes:
> Nothing of him that doth fade,
> but doth suffer a sea change
> into something rich and strange."
>
> W. Shakespeare, *The Tempest*, Act 1, scene 2.

1 Introduction

As made obvious by this workshop, there is a growing recognition of the relevance and usefulness of palaeohistology by scientists specialized in the fields of physical anthropology, forensic medicine, historic and prehistoric archaeology, archaeozoology, ethnology and many related sciences. Indeed, microscopic examination of actual tissue remains of ancient humans and animals (mostly hard tissues of the skeleton) can provide much useful information. It is relevant to, for example, individual age at death, palaeodemography, behaviour, available weapons and tools among historic and prehistoric human populations, as well as to seasonality, individual age, slaughtering and butchering practices and techniques, and perhaps domestication of vertebrates (mostly mammals) used or preyed upon by our ancestors.

For most current studies in these fields of research, the absolute age of the skeletal material used ranges from contemporary (that is, from the dissection room or forensic medicine material) to a few centuries old, or at most to about 10000 years B.P. This would take into account the complete realm of past "classical" human civilizations recorded by history down to their roots in the prehistoric "neolithic revolution". Although long, from an individual human perspective, it should be emphasized that this time scale is hardly significant compared to the geological time scale currently used by evolutionary palaeontologists.

Accordingly, vertebrate palaeontologists specialized in palaeohistology will be generally involved in researches which are rather distinct in their aim and scope to the ones conducted by archaeologists or anthropologists, even if many technical or methodological problems, such as those dealing with structural, functional and taxonomic interpretations of hard tissues, are obviously shared by both.

A case in point, when contrasting the approaches to palaeohistology by archaeologists and palaeontologists respectively, is the nature of the material available. While the hard tissues of palaeozoic or mesozoic age dealt with by palaeontologists are actually fossilized (one could almost say by definition), it is worth stressing that most of the anthropological or archaeological material is not, nor would it generally become fossilized in the usual sense of the word under normal circumstances. Hence, such material routinely shows evidence of a varying extent of post mortem degradation, whose study and interpretation becomes a very important subject of research by itself for specialists of "archaeological palaeohistology" (see, for example, Garland; Grupe & Dreses-Werringloer, both this Vol., Bell 1990 and references therein). Of course, the situation is not always as clear cut as this between the "archaeological" and "palaeontological" material, as it would be preposterous to set a priori an absolute time limit (say

10000 years) between "modern" material (still rich in organic components and prone to bacterial and mycelial destruction) on the one hand, and "fossil" material (where petrographic/geochemical processes such as diagenesis have a prevalent role) on the other hand.

While relatively modern material may sometimes experience "rapid" changes towards a truly fossilized condition (see, for example, Baud and Morgenthaler 1952, 1953), conversely, in some cases, fossils of very high absolute age (ranging in millions of years) may sometimes preserve a surprisingly high content of organic material (e.g. Bocherens 1990), such as the recently discovered Upper cretaceous dinosaurs from the northern slopes of Alaska (Davies 1987). Moreover, some of the post mortem alterations so prevalent in relatively recent archaeological bony material may sometimes be recognized in much older, truly fossilized palaeontological material, depending on the taphonomical circumstances to which it was submitted before ultimate fossilization (see, for example, Kiprijanoff 1881-83; Schaffer 1889).

Such problems point towards the usefulness of defining, in terms of physics and chemistry, what indeed is bone (and other hard tissue) fossilization, and to explore the morphological, histological and taphonomical correlation and consequences of this process (see, for example, Rogers 1924; Paine 1937; Baud and Morgenthaler 1952, 1953 and 1954; Mosebach 1974; Behrensmeyer and Hill 1980; Shipman 1981; Glimcher et al. 1990).

The purpose of the present chapter is to offer a survey of <u>some</u> of the topics of vertebrate palaeohistology as perceived from the points of view of the vertebrate zoologist, palaeontologist and comparative histologist as students of overall vertebrate evolution, in contrast to the points of view of the archaeologist, anthropologist or forensic medicine specialist. Indeed, there are different palaeohistological traditions rooted in these distinct fields, as stressed above, even if they share, of course, several subjects and approaches in common.

2 Historical Background

Vertebrate comparative palaeohistology could not begin, on the one hand, without the availability of some knowledge about the hard tissue histology of extant vertebrates including man, and, on the other hand, without the development of the technical know-how in order to process thin sections of fossil material. These conditions were met during the early part of the last century (see Enlow 1963; Francillon-Vieillot et al. 1990) and the starting point of vertebrate comparative palaeohistology may be traced back to the renowned work of Agassiz *Recherches sur les poissons fossiles*, published in Neuchatel (1833-1844) (see Dupuis 1988). Early works relied on polished sections of fossils, but true thin sections, allowing observations with the transmission compound microscope were in common use by the middle of the last century. The use of polarized light was pioneered by Valentin (1861) and von Ebner (1875). While thin section techniques became a routine tool for such fields as palaeobotany, micropalaeontology and invertebrate palaeontology, they have been used more rarely and somehow more reluctantly by vertebrate palaeontologists then and even up to the present day, with the exception of "lower" vertebrate specialists, who have built up an impressive and uninterrupted tradition in palaeohistology from Agassiz up to now. The explanation of this situation may be understood, retrospectively, for the following reasons. Firstly, fossils of "higher" vertebrates (i.e. mammals) were often rare and most valuable material, and it is understandable that curators of collections were (and still are) reluctant to submit such vertebrate fossils to destructive techniques. Second, in contrast to the meaningful information gained from the gross morphological study of "higher" vertebrate fossils thanks to the methods of comparative anatomy, scientific results gained from early histological investigations of fossil vertebrates were restricted in scope and interest. The situation was thus different for vertebrate and other palaeontologists, for whom the frequent paucity of meaningful morphological characters at the gross anatomical level caused a strong incentive towards palaeohistological (microstructural) studies by thin sections, especially in palaeobotany. The limitation of interesting results gained from early palaeohistological studies of, especially higher, vertebrates was caused by similar limitations in the basic knowledge and understanding linked to the then current state of the art in interpreting modern tissue histology and biology. Third, and more specifically, by training and tradition, the kind of questions addressed at histological investigations by late nineteenth century palaeontologists were first and foremost taxonomic, rather than functional, in scope: a somewhat deceptive approach to vertebrate palaeohistology, as shall be seen later. Indeed, the insistence on using hard tissue palaeohistology (and bone comparative histology among extant species) as a taxonomic-systematic tool rather than as a (palaeo) biological "record" has plagued this field for decades (see comments by Enlow 1963 & 1966; de Ricqlès 1975-1978; 1986; de Ricqlès et al. 1991). This is not to say, of course, that there are no means to extract sound systematic-phylogenetic information from vertebrate palaeohistological data (see below), but the logic, prospects and practice of such endeavours require a careful and specific evaluation.

For practical purposes, a summarized historical survey of vertebrate palaeohistological publications may be divided into distinct areas of research subject:

1. The dermal skeleton, especially among "lower" (non-tetrapod) vertebrates, which contain bony <u>and</u> dental tissues combined;
2. The endoskeleton among "non tetrapod" vertebrates;

3. Dermal bones and flat bones of tetrapods;
4. Compact bone of tetrapods, mainly studied on shaft cross-section;
5. Long bone growth and endochondral ossification;
6. Tooth tissue and associated tissues.

Finally, comparative histological works dealing with similar topics among extant vertebrates have been, and still are, developed in parallel with palaeohistological researches and prove invaluable for the proper interpretation of fossil material. In each of the above sections, the emphasis of research works may have been more or less put on, for example, comparative, structural, developmental, functional (biomechanical) or systematic, phylogenetic-evolutionary or ecological issues.

1. The dermal skeleton of agnathans, acanthodians, placoderms, chondrichthyans, actinopterygians and sarcopterygians include scales, spines, dermal rays and flat bones often composed of osseous tissues closely associated with dental tissues and showing a quite extensive spectrum of tissue diversity. Early palaeohistological investigations in this field include, for example, Agassiz (1833-44), Williamson (1849, 1851), Tomes (1878, 1898), Walcott (1892) and Goodrich (1904, 1907, 1913) who was among the first to use explicitly the term "paläohistology" in the title of one of his studies (Goodrich 1913). These works have been conducted in close connection with parallel researches in comparative histology, as published by, for example von Kölliker (1859), Baudelot (1873), Hertwig (1874a, 1876, 1879, 1882), Klaatsch (1890), Tretjakoff and Chinkus (1927, 1930) and many others.

More recent classical palaeohistological investigations in this extensive area of research include works by Aldinger (1937), Andrews (1973), Bryant (1936), Denison (e.g. 1963, among numerous publications), Gross (1935, 1956, 1961, 1967), Halstead-Tarlo (1964-65, 1973), Kerr (1952), Meunier (1984), Meinke and Thomson (1983), Ørvig (1957, 1965, 1968, 1977, 1978a, b, c), Reif (1982), Schaeffer (1977), Schultze (1966, 1969, 1977) and Thomson (1975); numerous contributions have also been published during the last 10 years or so.

2. Palaeohistological studies on the endoskeleton of non tetrapod vertebrates have not been as numerous and extensive as those dealing with the dermal skeleton and teeth. One may mention here the comparative histological studies by Stephan (1900), Haines (1934, 1937) and Wurmbach (1932) of extant fishes, and above all the masterful palaeohistological study by Ørvig (1951) on the endoskeleton of placoderms and fossil elasmobranchs, which offers a complete survey of the older literature. More recently, specific onto-histogenetic or palaeohistological researches on the endoskeleton of "fishes" have been published by Amprino and Godina (1956), Baud and Morgenthaler (1953), Ørvig (1957, 1965), Francillon (1974, 1977), Kemp and Westrin (1979), Bordat (1987) and Meunier (1987).

3. Dermal bones, flat bones and osteoderms of tetrapods: this somewhat heterogeneous category contains elements of the endoskeleton (such as "flat" bones from the limb girdles) and dermal bones, notably dentigerous jaw bones, as well as osteoderms and bony scales. Dermal bones and osteoderms of tetrapods often show superficial "ornamentations". Most palaeohistological knowledge on this material derives from the works of, for example, Credner (1893), Gross (1934) and, notably, Bystrow (1935, 1938a, 1944, 1947), supplemented by research on fossil osteoderms and scales by, for example, Colbert (1955), Findlay (1968, 1970), Schaeffer (1977) and Tchudinov (1970). Comparative histological surveys of similar mate-

rials among extant vertebrates have been provided by Otto (1908), Foote (1928), Peyer (1931), Enlow (1963, 1968), Castanet (1974) and de Buffrenil (1982). The general issue of the comparative significance of mineralized scales and osteoderms among tetrapods has been most recently reviewed by Zylberberg and Wake (1990) in their analysis of the fine structures of apodans (gymnophion) scales.

4. The long bones, as best exemplified by the endoskeletal limb bones of tetrapods, have provided the main basis for comparative histology of compact bone tissues, as found in diaphyseal cross-sections. Proeminal comparative palaeohistological descriptions of compact bone tissues among fossil amphibians, reptiles, birds and mammals have been provided by, for example, Queckett (1855), Kiprijannoff (1881-83), Schaffer (1899), Seitz (1907), von Nopcsa and Heidsieck (1933, 1934), Gross (1934), von Eggeling (1938), Enlow and Brown (1956-58), Peabody (1961), Currey (1960, 1962) and Enlow (1969), and summarized by de Ricqlès (1975b). They were matched by parallel efforts in comparative histology of extant vertebrates, with important contributions by Foote (1913, 1916) Matyas (1929), Weidenreich (1930), Petersen (1930), Amprino and Godina (1947) and Smith (1960). More recently, the renewed efforts at describing and interpreting functionally the compact bone histological diversity among fossil tetrapods have been somewhat influenced by the problem of the evolution of metabolic/thermal physiology, notably among dinosaurs (e.g. de Ricqlès, 1974, 1975, 1980; Reid 1981, 1983, 1984, 1985). On the other hand, the recognition of the practical possibilities

---

Additional references dealing with current research rather than with a historical overview of the development of comparative palaeohistology may be found in, for example, Francillon-Vieillot et al. 1990; de Ricqlès et al. 1991).

of skeletochronology to decipher individual age and demographic issues among living (and fossil) vertebrates has prompted an explosion of comparative histological data on bone (and cementum) of mammals, amphibians and reptiles [see, for example, reviews by Castanet (1985, 1986-87); Perrin and Mirrick (1980), Grue and Jensen (1979), Francillon-Vieillot et al. (1990)].

5. In retrospect, it is surprising how the problems of long bone lengthening during growth via the process of endochondral ossification in the epiphy-metaphyseal regions have been studied rather independently from compact (diaphyseal) bone. However, both are distinct, but clearly related, expressions of the general issue of bone growth modelling and remodelling dynamics. Famous comparative histological surveys of epiphyses and endochondral ossification among tetrapods include those of Moodie (1908), von Eggeling (1911, 1938), Froböse (1927), Heidsieck (1928, 1929), Lubosch (1927) and Roumiantsiev (1958). However, the most important comparative syntheses on the question have been provided by Haines (1938, 1942, 1969), and recent additions include reviews by Francillon (1981), de Ricqlès (1979), Rhodin (1985)and Buffrenil et al. (1987).

During the last century, the problem of endochondral ossification was dominated by lengthy discussions over the famous controversy between addicts of chondro-osseous metaplasia versus neoplasia, a debate traditionally settled by Müller's (1858) contribution. However, more recent surveys - (Haines and Mohuiddin 1968) and especially Beresford (1981) - as well as current researches in developmental biology (Hall 1975, 1978; Huysseune and Verraes 1986, 1990) indicate how complex the issue is.

Specific palaeohistological data on chondro-osseous relationships among fossil tetrapods include, for example, Hasse (1878), Ørvig T (1953), de Ricqlès (1975, 1989), de Buffrenil et al. (1987, 1990) and

Reid (1984), sometimes in connection with issues of developmental heterochronies.

6. Extensive palaeohistological and comparative descriptions of tooth tissues among fossil and extant "fishes" and tetrapods began with the renowned survey by Owen (1840-45). In view of the extreme polymorphism of dental tissues among lower vertebrates, an exceptionally high number of papers have been devoted to this field of study. Hertwig (1874a, b), Thomasset (1930) and Lison (1941), among many others, have provided valuable contributions. Similarly, numerous descriptions of labyrinthodont tooth structures among sarcopterygians and stegocephalians were published in the late 19th and early 20th centuries, and later synthetic palaeohistological descriptions of plicidentine and related tissues have been provided by, for example, Bystrow (1938b) and Schultze (1969). Peyer (1968), Schmidt and Keil (1971), Miles (1967) and most recently Lester and Von Koenigswald (1989), Carlson (1990) and Smith (1988) have offered extensive comparative surveys of the terminological, structural, functional and evolutionary problems linked to the various dental tissues in all the vertebrate classes, including a thorough analysis of all the palaeohistological data. The above mentioned list of papers cannot, however, give an idea of the massive amount of comparative histological literature dealing with fossil and extant dental tissues.

3 Causes and Taxonomic Consequences of Bone Histological Diversity

3.1 General

In the context of taxonomic determination of bone specimens of unknown pedigree, several comparative and palaeohistological studies as noted above, have been attempted. This research program is legitimate, but it has always been hampered by the fact that bone tissues generally show an extensive typological-structural diversity even at the level of the individual. Bone histology records, first and foremost, the local conditions of skeletal deposition and growth during ontogeny and, also, the functions during later life. Generally, those conditions and functions are different between various fields in a thin section, various parts of a given bone, various bones within an individual skeleton and at individual ages for all bones. Such various local biological circumstances produce locally diverse bone histological patterns which constitute a significant "record" of the said circumstances, a "record" which may be deciphered. A comparison with stratigraphy, which allows a "reconstruction" of the local history of a landscape in geology is obvious.

The resulting histodiversity has, thus, a complex, multifactorial causality which is mainly of an ontogenetic, functional origin, but which cannot generally be subsumed to any unique, single efficient cause (de Ricqlès 1975-78, 1980). This kind of functional causality of bone tissue diversity seems much more obvious and important than other more mediate or general causes, such as, for example, the "formal" taxonomic or phylogenetic origin of the specimen under examination, to account for its tissue structures (see Enlow 1966; de Ricqlès 1986; Francillon-Vieillot et al. 1990, for further discussion).

For instance, one could select from the skeleton of virtually any tetrapod vertebrate, local regions which would show, relative to others, either very similar or very different bone histological phenotypes. Such tissue variations would merely express similarities or differences in the local ontogenetic (growth) patterns used to build up the local bone morphology within a required, specific amount of time. Since similar circumstances of bone growth and remodelling can be met in various parts of the skeletons of taxonomically unrelated species (especially if they share convergent adaptations in terms of size, growth rates, ecology etc.) similar tissue phenotypes will be found in them, irrespective of phylogeny.

These data provide the rationale from which any attempt at a significant use of comparative bone histology for taxonomic purposes, and especially systematic determination of fossil bone samples of unknown pedigree, can be performed (Enlow 1966).

3.2 Some Speculations on the Relationships Between Bone Tissue Phenotypes and Genomes

As summarized in the previous section, comparative histological and ontogenetic data suggest that bone tissue phenotype diversity is the result of a complex and multifactorial determinism (see de Ricqlès 1975-78, 1980). Before exploring further the consequences of these data on the usefulness of bone histology as a taxonomical tool, some remarks on the possible genetic background of bone histological diversity may be in order.

On the problem of which structural level of organization is really "encoded" in the genome, it may be that if general bone forms and shapes are somehow controlled genetically, this may perhaps not be the case for the histological patterns, or tissue types, which actually make up these bones. In other words, it could be that no genes exist which would uniquely and directly specify lamellar or plexiform or any other compact bone tissue type itself. As previously discussed (de Ricqlès 1979), this conclusion is suggested by the great discrepancies in bone histology observed between two conspecific individuals (or two closely related species) with very distinct growth rates (Amprino 1947). It is also supported by the amazing amount of histological homoplasy (parallelism or convergence) of primary compact bone tissues in widely unrelated vertebrate species sharing some significant adaptations (for example, large sizes and high growth rates).

The differentiation of a well defined bone tissue phenotype (for example, plexiform bone tissue, as found in some artiodactyl mammals, some dinosaurs and some mammal-like reptiles, etc.) directly controlled or specified by the same gene(s) in these various vertebrate groups, acquired independently through the usually accepted mechanisms of chance point mutations selected at the phenotype level within populations, seems highly unlikely. The extreme locally specific histological differences between bone tissue phenotypes within any individual skeleton also raise difficulties for the hypothesis of a direct, specific genetic control for each bone tissue phenotype, in view of their diversity, number, and local transitions from one type to another.

What, thus, seems more likely to be directly controlled genetically would not be the actual bone tissue phenotype itself, but rather the local "morphogenetic program" (for example, amount of new primary bone to be deposited locally at a given rate, etc.), a concept sometimes also expressed as a "growth field" (Enlow 1968), in fact, a full array of interconnected growth and remodelling fields, which may not coincide with the actual anatomical limits of individual bones (see Bystrow 1935; Enlow 1968). In this way, every time that, for example, a high growth rate were to be locally specified (whatever the genetic encoding of the "morphogenetic program", which need not be homologous among various taxa or classes) similar phenotypical responses at the tissue level would arise in the form of specific (for example plexiform) tissue phenotypes. This could be causally explained at the histological level since all the structural details of, for example, plexiform tissue and the way it is laid down can be readily understood physiologically, in terms of local adaptation to a high rate of new bone deposition. At the same time, the various epigenetic constraints, notably biomechanical factors, which so obviously modulate the activities of bone depositing cells would also play a role in the final result, in the actual tissue type laid down locally, and in its subsequent fate (erosion, reconstruction).

Conversely, one could consider the hypothesis that the various bone tissue phenotypes are indeed direct expressions of specific genes. In that case, similar bone histological patterns (for example, plexiform tissue) found in different, unrelated taxa, are specified by the same gene(s). However, if it was so, one has to explain, as noted above, the peculiar taxonomic repartition of bone tissue phenotypes among various discrete lineages and in taxa sharing analogous adaptations (size, growth rates ecology etc.). If one thinks it is unlikely that similar genes evolved independently among various lineages (see above), then the most parsimonious interpretation would perhaps be plesiomorphy, i.e. passive conservation, at least in most tetrapod clades, of very ancient common genes able to control the various bone tissue phenotypes.

Some of these genes (for example, those specifying plexiform tissue) would be triggered each time fast, massive primary bone deposition was selected. In such cases, one can note that each phenotypically similar bone tissue type would be truly homologous between systematic groups, rather than only expressing homoplasies. A problem with this hypothesis would perhaps be the passive conservation, without chance drift, of potentially functional genes not submitted, sometimes for millions of years in some lineages, to phenotypic expressions and, hence, control by natural selection. For instance, in the synapsid clade the gene(s) that would control deposition of laminar and plexiform bone tissues, as expressed respectively among Upper Permian dycinodonts and Triassic kannemeyerids would be left "silent" for about 170 million years, before expressing themselves again phenotypically among large Cenozoic ungulates.

In summary, even if one assumes the possibility of a direct relationship between gene(s) and bone tissue phenotypes the actual taxonomic repartition of bone tissue types, when matched to current cladograms of vertebrates, and notably tetrapods (Gauthier et al.

1988), strongly suggests the occurrence of numerous histological convergences (homoplasies) between distantly related taxa sharing analogous adaptations. Alternatively, this repartition could be caused in part by histological homologies between tissue types via plesiomorphic conservation of ancestral homologous genes.

Hence, the phenotypic characteristics expressed in bone tissue diversity would often represent either homoplastic or plesiomorphic conditions. Both situations are obviously not conducive to a significant use of these characters in phylogenetic (and hence taxonomic) studies, as is demonstrated by cladistic methodology. This, in turn, may help us understand why so many attempts, over the last 150 years, at using bone histology as a tool for taxonomic determination of unknown bone specimens has often met with deceptive, doubtful or, at the very best, partial success.

3.3 Possible Consequences of the Determinism of Bone Tissue Polymorphism on the Use of Bone Histology as a Taxonomic Tool

As discussed in preceding sections, an important part of bone comparative histology and palaeohistology has always been the possibility of using histological characters for the identification of bone samples, a goal which cannot be described as having been successfully met. In retrospect, this situation may appear not only quite logical, but may also point to the prospects by which comparative histology of bone may prove really useful in the fields of comparative, developmental and evolutionary biology.

According to one of the interpretations mentioned, bone as a tissue would not be a direct expression of a genomic determinism, but rather a very indirect phenotypic reflection of some aspects of the genotype. Bone tissue phenotypes would be strongly mediated via the "canals of ontogenesis" including topological and constructional constraints of epigenetic origin (Waddington 1957; Alberch 1982). If this was so, it would not be surprising that comparative bone histology could neither be generally used as direct prima fascie evidence of the taxonomic (phylogenetic) situation of a given taxon, nor express the general patterns of irreversible law-like, cumulative changes usually ascribed to actual evolutionary changes (defined as direct phenotypic expressions of genomic changes). Note that this situation may differ with other hard tissues, such as enamel (Boyde 1978; Koenigswald 1989; Carlson, 1990).

Conversely, the other hypothesis, namely that bone tissue phenotypes are a direct expression of taxa-specific genes would have quite distinct consequences. If bone tissue phenotypes were the direct expression of clade-specific (apomorphic) genes, it would generally be natural to find bone histological characters quite diagnostic, phenotypically, of taxa, clades or lineages, as is the case with, for example, so many biochemical and gross anatomical characters. This assumption has been (implicitly) at the basis of many comparative histological studies of the past, which have attempted to demonstrate lineage or taxa-specific histological patterns and their evolution. For instance, a logical (if naive) endeavour, within the framework of such a research program, would be to seek bone histological characters pertaining, in a diagnostic way, to the synapsid clade, in the same way as the low temporal fenestra is a diagnostic anatomical "trade mark" (or autopomorphy) of this lineage. Indeed, precisely this research program was attempted many years ago by myself!

How do the histological data fit the two hypotheses? The actual segregation of bone tissue types, as described by available palaeohistological surveys among various vertebrate clades, do in no way falsify the first hypothesis. Bone tissue phenotypes are anatomically and taxonomically segregated in all vertebrates for reasons which, to all intents and purposes, can be first and foremost interpreted in terms of local (anatomical) ontogeny and functions (Gross 1934;

Amprino 1947; Enlow 1963, 1966). There are no obvious overall taxonomic reasons behind the general or local segregation of bone tissue types. There are no obvious "phylogenetic" or "evolutionary" trends either among bone tissues from early "primitive" towards later "advanced" condition(s). Man's dense Haversian bone tissue is not, as often assumed, more "advanced" or "evolved" than any other compact bone tissue type.

But the second hypothesis is not completely falsified by the evidence either. It is true that some bone histological characteristics may sometimes appear as taxa-specific and, if they are not caused by homoplasy, used, at least tentatively, as synapomorphies. Aspidin of heterostracans, avascular bone of squamates, acellular bone of "higher" teleosts and bone with Willamson's canaliculi of holosteans could be taken as tentative examples, among others, and many more may be discovered in the future using finer quantitative histological surveys, as permitted by image analysis. Such cases would merely emphasize the rather trivial fact that, as any component of the body at any level of integration, bone tissues are also ultimately a phenotypic expression of the genome itself, perhaps in such cases only less intensively modulated and mediated by local ontogenetic/epigenetic constraints than among some other vertebrate clades.

From another point of view, such examples of an agreement between bone histology and supra specific taxa (presumably monophyletic or clades) raises the issue of a proper use of bone histological characters in a cladistic analysis of vertebrates (Maisey 1988); see also Carlson (1990) for a similar approach to enamel. For instance, it could be assumed that in the endoskeleton, compact cellular bone of periosteal origin with longitudinal primary vascular canals and well expressed growth cycles can be taken as the plesiomorphic condition for bony vertebrates. This condition, on the other hand, may be taken as an apomorphy of osteichthyans in a cladistic sense (viz. including actinopterygians, sarcopterygians and tetrapods in the usual sense).

4. Concluding Remarks and an Agenda for Bone Palaeohistology

In view of the common interest in vertebrate hard tissue histology shown by many scientists with very distinct training, traditions and research goals, from molecular and cell biologists to evolutionary palaeontologists via orthopaedic surgeons or archaeologists, it is not surprising that difficulties in mutual understanding may arise. Hence the following suggestions, drawn from personal training (and prejudices) in comparative and evolutionary palaeohistology are only purported in order to stimulate reactions from those trained within other traditions.

4.1 On Terminology. Typological and Functional Descriptions of Bone (and other) Hard Tissues.

Gross misunderstandings (at least as far as bone tissue is involved) stemmed in the past from the non-standardized and improper usage of terms (hence, also of concepts) in bone histology. Terms such as, for example, "Haversian bone", primary versus secondary bone, laminar versus lamellar bone, etc., have been used and misused in numerous, contradictory ways, a situation which has considerably blurred (and it still does) sound interpretation of many histological studies. The current availability of image analysis is a strong incentive towards standardizing of bone terminology for bone morphometry and studies in densitometry (e.g. Recker 1983). Based on an indepth analysis of the comparative literature, the diversity of bone tissue patterns, as well as related functional and terminological problems have been extensively reviewed and discussed (Enlow 1966; de Ricqlès 1975-78, 1975b; Francillon-Vieillot et al. 1990; Reid 1983,

1984, 1985) and it is hoped that such overviews of comparative bone histological terminology may help towards a standardization of vocabulary, one which at least would help to prevent further mis-understandings over words.

4.2 Too Much Emphasis on Secondary Bone?

Because they deal with human tissues and pathologies, a very large part of the activities of histologists and bone biologists within the biomedical and archaeological fields is focused on the problems of bone remodelling (that is, erosion and reconstruction, and bone turn-over) in spongy and compact bone tissues. Trabecular bone adaptation to strain or osteoporosis in spongy bone and Haversian reconstruction in compact bone pervade most current biological studies of bone as a tissue and dense Haversian bone tissue is still described in textbooks of histology as **the** typical compact bone tissue. A wide comparative assessment of bone histology, however, has revealed for a long time the extreme taxonomic prevalence and great diversity of primary (not reconstructed) compact bone tissues (see Fig. 1). Compact primary bone tissues of periosteal or dermal origin offers a wide spectrum of histological variation which has a most important comparative as well as histophysiological significance (Amprino 1947; Enlow 1966; de Ricqlès 1975-78) since it expresses, for example, the local dynamics of bone deposition.

Only a tiny fraction of the efforts devoted so far by physiologists, endocrinologists and biochemists towards the elucidation of the systemic and local control of bone remodelling, would be likely to lead to most interesting breakthroughs in bone histophysiology, if devoted to the elucidation of the fine biological control of primary bone histodiversity.

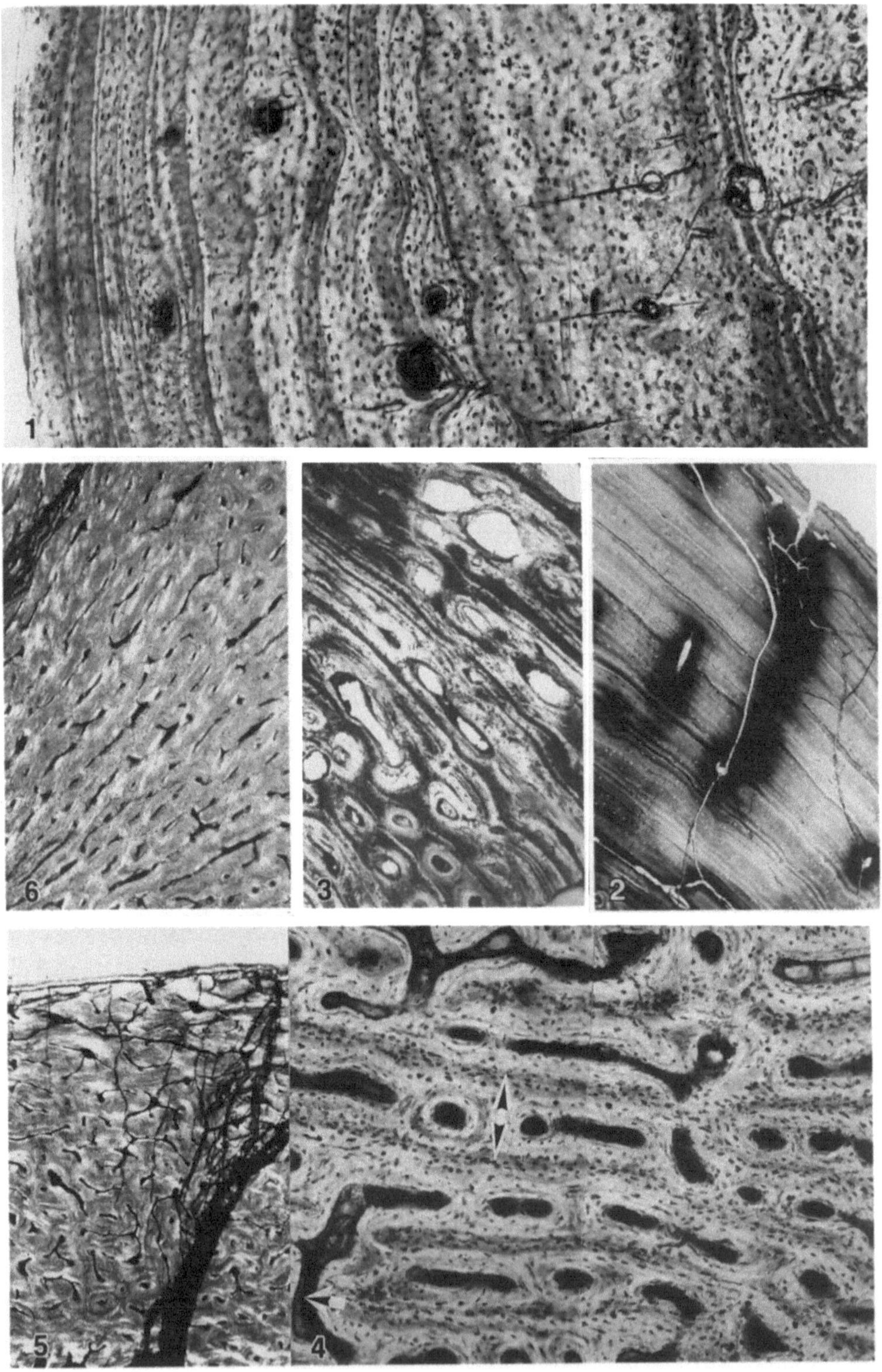
1
6
3
2
5
4

Fig. 1-6 Primary (periosteal) bone histodiversity

Compact periosteal (appositional) bone tissue forming long bone shafts in tetrapods can be classified into various typological (and functional) categories, according to the fibrillar and vascular patterns. "Lamellar-zonal" bone tissues (Figs. 1-3) have a finely or coarsely lamellated matrix, or a parallel-fibred (pseudo lamellar) matrix, deposited centrifugally by the periosteum. Vascularization density varies greatly. These tissues generally indicate low to moderate (cyclical) growth rates.

The "fibro-lamellar complex", on the other hand (Figs. 4-6) lumps tissues which are formed by the close association of (1) a more or less coarsely woven fine cancellous periosteal bone matrix with (2) lamellar bone matrix deposited centripetally as numerous primary osteons in the cavities of the woven bone. This tissue complex varies typologically according to its vascular pattern and generally indicates high to very high (continuous) local growth rates. A continuum of intermediate patterns can link all the various typological categories recognized, expressing intermediate conditions in speed of deposition and cyclical versus continuous deposition dynamics.

1. Detail of lamellar-zonal tissue with extensive evidence of growth cycles in the rib of *Mesosaurus*, a small aquatic reptile from the Lower Permian of Brazil. Scattered small, simple, primary, longitudinal vascular canals can be seen. External cortex cross section, viewed by ordinary light microscope

2. Overall view of lamellar-zonal tissue in *Claudiosaurus*, a small Upper Permian reptile of uncertain pedigree from Madagascar. Numerous growth cycles in the poorly vascularized cortex are obvious. Bone tissue is very similar to that observed in many extant small amphibians and reptiles. Cortex of a long bone shaft, cross section, viewed by ordinary light microscope (Courtesy of Dr. V. de Buffrénil)

3. Overall view of well vascularized lamellar-zonal tissue in a large aquatic stegocephalian (undetermined capitosaurid from the Triassic of Sahara). The zones, well vascularized by numerous primary osteons, regularly alternate with avascular annuli formed by grossly lamellar periosteal bone. External cortex, viewed by ordinary light microscope

4. Detail of the "fibro-lamellar complex", here in the form of the so-called laminar bone tissue pattern. The tissue is formed of superimposed circumferential laminae (double headed arrow). Woven periosteal bone matrix is darker and more cellular than lamellar osteonal material laid down around the longitudinal and circular vascular canals. Occurrence of a few radial vascular canals (single headed arrow) gives a "subplexiform" pattern of vascularization to the tissue. This bone tissue type is found among many large extant mammals and birds and indicates high to very high local growth rates. Cross-section from a long bone shaft of a dicynodont, a plant eating mammalian-reptile (Upper Permian of South Africa), viewed by ordinary light microscope

5. "Fibro lamellar complex": overall view. Here, the irregular spatial organization of the numerous primary osteons results in the so-called reticular bone tissue pattern. Long bone shaft of the gorgonopsian *Aelurognathus*, a carnivorous mammalian reptile from the Upper Permian of South Africa; viewed by slightly polarized light

6. "Fibro lamellar complex": overall view. The prevalence of radial vascular canals over ,longitudinal and especially circular, canals results here in the formation of the so-called radiating bone tissue pattern. Long bone shaft of a whaitsiid therocephalian, a carnivorous mammalian reptile from the Upper Permian of South Africa; viewed by slightly polarised light

4.3 A Healthy Palaeopathology

It seems that there is an intense fascination for this field of research shared by many anthropologists and archaeologists. However, in order to develop, palaeopathology may have to eradicate some epistemological problems. Since its origins, this field had some obvious, but sometimes debatable, connections with comparative histology, palaeohistology and palaeontology, as evidenced by early classical works of, for example, Abel (1912), Moodie (1923) and Pales (1930), and now by, for example, Rotschild (1987a, b, 1988). It seems to the present writer that the study of data on the antiquity and possible evolution of injuries and disease, as far as they can be recorded by palaeozoic or recent skeletons, is a legitimate but highly specialised field of palaeobiology which raises specific and extremely difficult methodological problems. In such a domain, the study and knowledge of "normality", that is, healthy normal conditions should always be the previous standard of comparison and a priori palaeopathological interpretations may not be warranted. Indeed, from a situation when "obvious" palaeopathological conditions (for example, an exceptional vertebral synostosis in a dinosaur) may be accepted, there is the possibility to "creep" towards wider and wider "palaeopathological" interpretations of fossil diversity (Abel 1912; von Nopcsa 1923, 1930; Kayser 1960, 1970) where most anatomical or histological characteristics of unfamiliar, extinct vertebrates may be regarded as pathologies of some sorts.

Apart from inherent anthropocentrism, this viewpoint also contributes to highly doubtful general considerations on evolutionary mechanics (orthogenesis, hyperthelism, typolysis, etc.; see Gould 1973, 1974). In this domain it seems that the adequate version of Occam's razor would be to reject a priori any palaeopathological interpretation as long as one is not absolutely compelled to accept it.

4.4 Intensive versus Extensive Comparative Histological Surveys

Proemial comparative studies on bone histology, such as Seitz (1907), Gross (1934) and Enlow and Brown (1956-58) and other similar works, can be described as extensive. In view of the paucity of fossil material available for sectioning, anatomically varied and heterogenous bone fragments were often used for description in a wide spectrum of taxa. However, the later recognition of the paramount importance of ontogenetic or anatomical factors for a proper interpretation of bone histological diversity (Enlow 1963, 1966, 1969; de Ricqlès 1975-78) raised questions on the real significance of the wide comparative approaches. In fact, the recognition of the ontogenetic factors which control local and general diversity of primary compact bone (Amprino 1947) makes it possible now not only to understand the functional significance of age-specific, local/anatomical-specific and species-specific histological differences, but even to predict them (de Ricqlès and Bolt 1983). This new knowledge makes it more useful now to concentrate on the intensive comparative palaeohistological description of a few related species, with full documentation of ontogenetic and bone-specific changes in the whole skeleton. More precise and reliable palaeobiological conclusions on data such as individual age, growth rate, ontogenetic changes at the basis of evolutionary changes in morphology, etc., are more likely to be extracted now from such "intensive" studies than from additional "extensive" surveys of bone palaeodiversity.

However, as noted by Enlow, no similar intensive histological surveys as those suggested above in palaeohistology seem to be available for extant species, even man. Such documents, if available, would nevertheless be of great value in providing histological comparative standards for further comparisons and such a research program is encouraged.

4.5 On Bone Form, Growth and Evolution: Further Prospects for Palaeohistology

In every vertebrate, bone number, gross shapes and proportions, and to a large extent, sizes, are highly useful as precise species-specific characteristics. Thanks to the methods of comparative anatomy and functional morphology, they provide, therefore, valuable characteristics for taxonomic, phylogenetic, functional and evolutionary studies.

It is obvious that the information necessary to shape each bone with exquisite precision within the space and time of ontogeny should somehow be encoded as part of each species genome. Since genomes do change with time (over long periods) and bone shapes are consequences of genomic-based information, gross morphological changes of bones can be defined as a biological evolution in the usual sense, that is, at the phenotype level. In fact, and conversely, the actual differences recorded between shapes of homologous bones, and their ordering into a taxonomy of hierarchical monophyletic clusterings, remains one of the strongest and compelling pieces of evidence for vertebrate evolution itself. In addition to artefactual selection, rapidly increasing, though still scattered, evidence for all this is now provided by the number of skeletal diseases, or various bone abnormalities, presently known to have a direct genetic basis (Johnson 1986).

However, the genes code for proteins, and even the knowledge of the gene coding for the bone collagen, as well as for bone specific non-collagenous proteins, still tells us very little, if anything at all, on how the crucial problem of bone growth and form shaping by modelling and remodelling is actually controlled.

On the other hand, since bones are formed of hard tissues which impose quite peculiar and precise constructional constraints on growth processes, such as accretional and no intussusceptional growth, and because they record, by a form of stratigraphy, their own ontogenetic processes in their structures, bone and other hard tissues are especially useful and well adapted to tackle this basic problem common to developmental and evolutionary biology: the creation and evolution of forms.

This is especially true if one considers not only the possibilities of comparing directly skeletal ontogenies in various related extant species, the comparative or synchronic approach, but also by following how bone ontogenies do change along actual phylogenetic sequences among fossils (the diachronic approach; see, for example, de Ricqlès 1979; de Ricqlès and Bolt 1983).

Bone comparative histology and palaeohistology thus offer a quite unique and exciting research program. They allow us not only to follow or merely describe evolutionary changes as anatomical end-results, but also to precisely analyse the actual evolutionary mechanisms, as they act at the phenotype level through ontogenetic changes in phylogenetic series. This also offers a most valuable possibility of analysing the relationships between ontogeny and phylogeny and, hence, the issues of evolutionary mechanisms via developmental heterochronies.

5 Summary

In this chapter, the approaches to palaeohistology from the point of view of the evolutionary palaeontologist, archaeologist and anthropologist are contrasted. After an overview of the historical development of comparative palaeohistology, a discussion is provided on the ontogenetic and possible genetic basis of bone tissue phenotype diversity and the consequences on the possibilities of, or lack of, using bone histology as a taxonomic tool. Finally, some points of view on specific issues, such as bone tissue terminology, palaeopathology and the prospects for future research on bone palaeohistology, are suggested.

Acknowledgements The author wishes to express his special thanks to Dr. Neil Garland for his practical help towards completion of this paper.

References

Abel O (1912) Palaeobiologie der Wirbeltiere. Schweizerbart, Stuttgart

Agassiz JLR (1833-44) Recherches sur les poissons fossiles. Neuchâtel

Alberch P (1982) Developmental constraints in evolutionary processes. In: Bonner TJ (ed) Evolution and development. Springer, Berlin Heidelberg New York, pp 313-332

Aldinger H (1937) Permische Ganiodfische aus Ostgrönland. Medd Grön 102:1-392

Amprino R (1947) La structure du tissue osseux envisagée comme expression de differénces dans la vitesse de l'acroissement Arch Biol. 58:315-330

Amprino R, Bairati A (1936) Processi di reconstruzione e di rissorbimento nella sostanza compatta dellle ossa dell'uomo. Z Zellforsch 24:439-511

Amprino R, Bairati A (1938) La variazioni nella struttura dell'osso in relazione al eta e la loro importanze medico forenze. Arch Anthropol Crim 16:61-74

Amprino R, Godina G (1947) La struttura della ossa nei vertebrate. Ricerche comparative negli amfibi e negli amnioti. Comment Pont Acad Sci 11:329-467

Amprino R, Godina G (1956) Osservazioni sul rinnovamento strutturale dell'osso in pesci teleostei. Publ Staz Zool Napoli 28:62-71

Andrews M (1973) Interrelationships of crossopterygians. In: Greenwood PM, Miles RS, Patterson C (eds) Inter-relationships of fishes. Linn Soc London, and Academic Press, New York, London, pp 137-177

Baud CA, Morgenthaler PW (1952) Recherces sur l'ultrastructure de l'os humain fossile. Arch Suisses Anthropol gén 17:52-65

Baud CA, Morgenthaler PW (1953) Sur l'ultrastructure du tissu osseux d'un poisson téléostéén fossile: Dyplomystus brevissimus (Fam. Clupeidae). Bull Mus Hist Nat Paris 25:641-645

Baud CA, Durif S, Morgenthaler PW (1954) Recherches sur la structure cristalline de l'os humain fossile. Arch Suisses Anthropol gén 19:37-52

Baudelot ME (1873) Recherches sur la structure et le développement des écailles des poissons osseux. Arch Zool Exp Gén 2:87-244; 429-480

Behrensmeyer AK, Hill AP (1980) Fossils in the making. Univ Press, Chicago

Bell LS (1990) Palaeopathology and diagenesis: an SEM evaluation of structural changes using backscattered electron imaging. J Arch Sci 17:85-102

Beresford WA (1981) Chondroid bone, secondary cartilage and metaplasia. Urban & Scharzenberg, Munich Baltimore.

Bocherens (1990) Mis en évidence du régime alimentaire végétarian de l'ours des caveunes (Ursus spelaeus) par la biogéochemie isotopique (^{13}C, ^{15}N) du collagène des vertébrés fossiles. CR Acad Sci Paris 331:1279-1284

Bordat C (1987) Etude ultrastructurale de l'os des vertèbres du sélacien Scyliorhinus canicula L. Can J Zool 65:1435-1444

Boyde A (1978) Development of the structure of the enamel of the incisor teeth in the three classical suborder groups of rodentia. In: Butler PM, Joysey KA (eds) Development, function and evolution of teeth. Academic Press, New York London, pp 43-48

Bryant WL (1936) A study of the oldest known vertebrates, Astraspis and Eryptichius. Proc Am Philos Soc 76:409-427

Buffrenil V de (1982) Morphogenesis of bone ornamentation in extant and extinct crocodilians. Zoomorphology 99:155-166

Buffrenil V de, Mazin JM, de Ricqlès A (1987) Caractères structuraux et mode de croissance du fémur d'Omphalosaurus nisseri, icthyosaurien du trias moyen des Spitzberg. Ann Paléontol 73:195-216

Buffrenil V de, Ricqlès A de, Ray CE, Domning DP (1990) Bone histology of the ribs of archaeocetes (Mammalia, Cetacea). J Vertebr Palaeontol 10:455-466

Bystrow AP (1935) Morphologische Untersuchungen der Deckknochen des Schädels der Wirbeltiere, 1. Mitt: Schädel der Stegocephalen. Acta Zool 16:64-141

Bystrow AP (1938a) Dvinosaurus als neotenische Form der Stegocephalen. Acta Zool 19:209-295

Bystrow AP (1938b) Zahnstruktur der Labyrinthodonten. Acta Zool 19: 387-425

Bystrow AP (1944) Kottlassia prima, Amatitzky. Bull Geol Soc Am 55:379-416

Bystrow AP (1947) Hydrophilous and xerophilous labyrinthodonts. Actal Zool 28:137-164

Carlson S (1990) Vertebrate dental structures. In: Carter JG (ed) Skeletal biomineralisation: patterns, processes and evolutionary trends, vol 1. Van Nostrand & Reinhold, New York, pp 531-556

Castanet J (1974) Etude histologique des marques squelettiques de croissance chez Vipera aspis L. Zool Scr 3:137-151

Castanet J (1985) La squelettochronologie chez les reptiles. I: Résultats expérimentaux. Ann Sci Nat Zool 7:223-40

Castanet J (1986-7) La squelettochronologie chez les Reptiles. III: Applications. Ann Sci Nat Zool 8:157-172

Colbert EH (1955) Scales in the permian amphibian Trimerochachis. Am Mus Novit 1740:1-17

Credner H (1893) Die Stegocephalen und Saurier aus dem Rothliegenden des Plauenschen Grundes bei Dresden. Dtsch Geol Ges Z 45:639-704

Currey JD (1960) Difference in the blood supply of bone of different histological types. Q J Microsc Sci 101:351-370

Currey JD (1962) The histology of the bone of a prosauropod dinosaur. J Paleontol 5:238-246

Davies KL (1987) Duck-bill dinosaurs (Hadrosauridae, Ornithischia) from the north slope of Alaska. J Paleontol 61:198-200

Denison RH (1963) The early history of the vertebrate calcified skeleton. Clin Orthop Rel Resch 31:141-152

Denison RH (1967) Ordovician vertebrates from western United States. Fieldiana Geol 16: 131-192

Denison RH (1974) The structure and evolution of teeth in lungfishes. Fieldiana Geol 33:31-58

Dupuis Cl (1988) Savoirs et idées à Neuchâtel, terroir séculairement fertile pour l'histoire naturelle. Bull Soc Neuchâtel Sci Nat 111:105-124

Ebner V von (1875) Über den feineren Bau der Knochensubstanz. Sitzungsber Kön Akad Wiss Math Naturwiss Wien III 72:49-138

Eggeling H von (1911) Der Aufbau der Skeletteile in den freien Gliedmassen der Wirbeltiere. Untersuchungen an urodelen Amphibien. Fischer, Jena

Eggeling H von (1938) Allgemeines über den Aufau knöcherner Skeletteile. In: Bolk L, Goppert E, Kalius E, Lubosch W (eds) Handbuch Vergleichende Anatomie der Wirbeltiere, vol 5. Urban & Schwarzenberg, Berlin, pp 275-304

Enlow DH (1963) Principles of bone remodeling. Thomas, Illinois

Enlow DH (1966) An evaluation of the use of bone histology in forensic medicine and anthropology. In: Evans G (ed) Int Congr Anat Symp Joints Bones. Springer, Berlin Heidelberg New York, pp 92-111

Enlow DH (1968) Wolff's law and the factors of architectonic circumstances. Am J Orthod 54:803-822

Enlow DH (1969) The bone of reptiles. In: Gans C (ed) Biology of the Reptilia, vol A.1. Academic Press, New York London, pp 45-80

Enlow DH, Brown SO (1956-58) A comparative histological study of fossil and recent bone tissues, pt 1-3. Texas J Sci 8:403-443; 9:185-214; 10:187-230

Findlay GH (1968) On the structure of the skin in Uranocentrodon (Rhinesuchus) senekalensis Van Hoepen. Paleontol Afr 11:15-21

Findlay GH (1970) Skin structure of small pareiasaurs. Paleontol Afr 13:15-23

Foote JS (1913) The comparative histology of the femur. Smithsonian Misc Collect 61 (Publ 2232)

Foote JS (1916) A contribution to the comparative histology of the femur. Smithsonian Contrib Knowl Publ 2382: pp 1-242

Foote JS (1928) Bone as a measure of development and when and why we acquired our teeth, pt 1-3. Opus posthumus. Douglas, Omaha

Francillon H (1974) Développement de la partie postérieure de la mandible de Salmo trutta fario. Zool Scr 3:41-51

Francillon H (1977) Développement de la partie antérieure de la mandible de Salmo trutta fario. Zool scr 6:245-251

Francillon H (1981) Contribution à l'étude des structures et du fonctionnement des épiphyses fémorales chez les amphibiens anoures. Arch Zool Exp Gen 122: 289-310

Francillon-Vieillot H, de Buffrénil V, Castanet J, Géraudie J, Meunier FJ, Sire JY, Zylberberg L, de Ricqlès A (1990) Microstructure and mineralisation of vertebrate skeletal tissues. In: Carter JG (ed) Skeletal biomineralisation: patterns, processes and evolutionary trends. Van Nostrand Reinhold, New York, pp 471-530

Froböse H (1927) Der Aufbau der skeletteile in den freien Gliedmassen der anuren Amphibien. Jahrb Morphol Mikrosk Anat 58: 473-566

Gauthier J, Kluge AG, Rowe, T (1988) Amniote phylogeny and the importance of fossils. Cladistics 4:105-209

Glimcher M, Cohen-Solal L, Kossiva D, Ricqlès A de (1990) Biochemical analysis of fossil enamel and dentine. Paleobiology 16:219-232

Goodrich ES (1904) On the dermal fin-rays of fishes- living and extinct. Q J Microsc Sci 47:465-521

Goodrich ES (1907) On the scales of fishes, living and extinct, and their importance in classification. Proc Zool Soc London 751-774

Goodrich ES (1913) On the structure of bone in fishes: a contribution to palaeohistology. Proc Zool Soc London 80-85

Gould SJ (1973) Positive allometry of antlers in the "Irish Elk"; Megaloceros giganteus. Nature (London) 244:375-376

Gould SJ (1974) The evolutionary significance of "bizzare" structures: antler size and skull size in the "Irish Elk" Megaloceros giganteus. Evolution 28:191-220

Gross W (1930) Die Fische des mittleren Old Süd-Livlands. Geol Paläontol 18: 121-156

Gross W (1934) Die Typen des mikroskopischen Knochenbaues bei fossilen Stegocephalen und Reptilien. Z Anat 103:731-764

Gross W (1935) Histologische studien am Aussenskelett fossiler Agnathen und Fische. Paleontographica 83A:1-60

Gross W (1956) Über Crossopterygier und Dipnoer aus dem Oberdevon im Zusammenhang einer vergleichenden Untersuchung des Porenkanalsystems paläozoischer Agnathen und Fische. Kungl Svensk Vetensk Akad Hand 5:1-140

Gross W (1961) Aufbau des Panzers obersilurischer Heterostraci und Osteostraci Norddeutschlands (Geschiebe) und Oesels. Acta Zool 42:73-150

Gross W (1967) Über Thelodontier-Schuppen. Paleontographica 127:1-67

Grue H, Jensen B (1979) Review of the formation of incremental lines in tooth cementum of terrestial mammals. Dan Rev Game biol 11:1-48

Haines RW (1934) Epiphysial growth in the branchial skeleton of fishes. Q J Micro Sci 77:79-97

Haines RW (1937) Posterior end of Meckel's cartilage and related ossification in bony fishes. Q J Micro Sci 80:1-38

Haines RW (1938) The primitive form of epiphysis in the long bones of tetrapods. J Anat 72:323-343

Haines RW (1942) The evolution of epiphyses and of endochondral bone. Biol Rev 174:267-292

Haines RW (1969) Epiphyseal and sesamoids. In: Gans C, Bellairs A (eds) Biology of the Reptilia. Academic Press, New York London, pp 81-115

Haines RW, Mohuiddin A (1968) Metaplastic bone. J Anat 103:527-538

Hall BK (1975) Evolutionary consequences of skeletal differentiation. Am Zool 15: 329-350

Hall BK (1978) Developmental and cellular skeletal biology. Academic Press, New York London

Halstead-Tarlo LB (1964-65) Psammosteiformes (Agnatha). A review with description of new material from the lower Devonian of Poland. Pt 1; and 2. Paleontol Pol 13:1-133; and 15:1-167

Halstead LB (1969) Calcified tissues in the earliest vertebrates. Calc Tissue Res 3:107-124

Halstead LB (1973) The heterostracan fishes. Biol Rev 48:279-332

Hasse C (1878) Die fossilen Wirbel. Die Histologie fossiler Reptilienwirbel. Morphol Jahrb 4:480-502

Heidsieck E (1928) Der Bau der Skeletteile der freien Extremitäten bei den Reptilien, 1. Mitt: Gekkonidae und Agamidae. Morphol Jahrb 59:343-492

Heidsieck E (1929) Der Bau der Skeletteile der freien Extremitäten bei den Reptilien, 2. Mitt: Hatteria (Sphenodon) functata. Morphol Jahrb 62:319-354

Hertwig O (1874a) Über Bau und Entwicklung der Placoidschuppen und der Zähne der Selachier. Jenaer Z Naturwiss 8:331-402

Hertwig O (1874b) Über das Zahnsystem der Amphibien und seine Bedeutung für die Genese des Skeletts der Mundhöhle. Arch Mikro Anat 11 (Suppl)

Hertwig O (1876) Über das Hautskelett der Fische. Morphol Jahrb Z Anat Entwickl 2:328-396

Hertwig O (1879) Über des Hautskelett der Fische. Morphol Jahrb Z Anat Entwickl 5:1-21

Hertwig O (1882) Über des Hautskelett der Fische. Morphol Jahrb Z Anat Entwickl 7:1-42

Huysseune A, Verraes W (1986) Chondroid bone on the upper pharyngeal jaws and neurocranial bones in the adult fish Astatotilapia elegans. Am J Anat 177:527-535

Huysseune A, Verraes W (1990) Histochimie des glucides dans l'os chondroide mûr chez Astatotilapia elegans. Ann Sci Nat Zool 11:29-43

Johnson DR (1986) The genetics of the skeleton. Clarendon, Oxford

Kaiser HE (1960) Untersuchungen zur vergleichenden Osteologie der fossilen und rezenten Pachyostosen. Paleontographica 114:113-196

Kaiser HE (1970) Das Abnorme in der Evolution. Acta Biotheoret Suppl 17. Brill, Leiden

Kemp NE, Westrin SK (1979) Ultrastructure of calcified cartilage in the endoskeletal tesserae of sharks. J Morphol 160:75-102

Kerr T (1952) The scales of the primitive living actinoterygians. Proc Zool Soc Lond 122:55-78

Kiprijanoff AV (1881-83) Studien über die fossilen Reptilien Russ lands. Mem Acad Imp Sci St Petersburg Ser 7: 28-31

Klaatsch H (1890) Zur Morphologie der Fischuppen und zur Geschichte der Hartsubstanzgewebe. Morphol Jahrb 16:97-203; and 209-258

Kölliker A von (1859) On the different types in the microstructure of the skeletons of osseous fish. Proc R Soc London 9:656-668

Kölliker A von (1889) Handbuch der Gewebelehre des Menschen. Engelmann, Leipzig

Lester K, von Koenigswald W (1989) Crystallite orientation discontinuities and the evolution of mammalian enamel- or when is a prism. Scann Microsc 3:645-663

Lison L (1941) Recherches sur la structure et l'histogenèse des dents des poissons dipneustes. Arch Biol 52:279-320

Lubosch W (1927) Das perennierende Kalkskelett der Wirbeltiere und der fibrillare Bau der knorpeligen Skelletteile. Z Mikro Anat Forsch 9:69-121

Maisey JG (1988) Phylogeny of early vertebrate skeletal induction and ossification patterns. Evol Biol 22:1-36

Matyas J (1929) Die mikroanatomischen Knochenstrukturen und die Abstammungsgeschichte einzelner Wirbeltiergruppen. In: 10th Congr Int Zool, Budapest, pp 66-74

Meinke DK, Thomson KS (1983) The distribution and significance of enamel and enameloid in the dermal skeleton of osteolepiform rhipidistian fishes. Paleobiology 9:138-149

Meunier FJ (1984) Spacial organisation and mineralisation of the basal plate of elasmoid scales in osteichthyans. Am Zool 24:953-964

Meunier FJ (1987) Os cellulaire, os acellulaire et tissus dérivés chez les osteichthyens: les phénomènes de l'acéllularisation et de la perte de minéralisation. Ann Biol 26:201-233

Miles AEW (ed) (1967) Structural and chemical organisation of teeth. Academic Press, London New York

Moodie RL (1908) Reptilian epiphyses. Am J Anat 7:443-467

Moodie RL (1923) Paleopathology. Univ Illinois Press, Urbana

Mosebach R (1974) A propos de la composition chimique et minérale de la substance inorganique des os et de la dentine de vertébrés actuels et fossiles, de sa structure et de sa groisseur du grain moyen en fonction de l'age gèologique. Bull Groupe Europ Rech Sci Stomat Odont 17:59-85

Müller H (1858) Über die Entwicklung der Knochensubstanz nebst Bemerkungen über den Bau rachitischer Knochen. Z Zool 9:147-233

Nopcsa F von (1923) Vorläufige Notiz über die Pachyostose und Osteosklerose einiger mariner Wirbeltiere. Anat Anz 56:353-359

Nopcsa Fr von (1930) Notes on Stegocephalia and Amphibia. Proc Zool Soc Lond 64:979-995

Nopcsa Fr von, Heidsieck E (1933) On the histology of the ribs in immature and half-grown trachodont dinosaurs. Proc Zool Soc Lond 1:221-226

Nopcsa Fr von, Heidsieck E (1934) Über eine pachyostotische Rippe auf der Kreide Rügens. Acta Zool 15:431-455

Ørvig T (1951) Histologic studies of placoderms and fossil elasmo branches I: The endoskeleton with remarks on the hard tissues of lower vertebrates. Ark Zool 2:321-456

Ørvig T (1953) On the mesozoic field of andoya. Acta Borealia A.4:1-31

Ørvig T (1957) Remarks on the vertebrate fauna of the lower Upper Devonian of Esumina bay, P.Q. Canada. Ark Zool 10:367-426

Ørvig T (1965) Paleohistological notes 2. Certain comments on the phyletic significance of acellular bone tissue in early lower vertebrates. Ark Zool 16:551-556

Ørvig T (ed) (1968) The dermal skeleton: general considerations. In: Current problems of lower tetrapods phylogeny. John Wiley & Sons, New York, pp 373-397

Ørvig T (1977) A survey of odontodes ("dermal teeth") from developmental, structural, functional and phyletic points of view. In: Andrews SM, Miles RS, Walker AD (eds) Problems in vertebrate evolution. Soc London Symp Ser 4: pp 53-75

Ørvig T (1978a) Microstructure and growth of the dermal skeleton in fossil actinopterygian fishes: Birgeria and Scanilepis. Zool Scr 7:33-56

Ørvig T (1978b) Microstructure and growth of the dermal skeleton in fossil actinopterygian fishes: Boreosomus, Plegmolepis and Gyrolepis. Zool Scr 7:125-144

Ørvig T (1978c) Microstructure and growth of the dermal skeleton in fossil actinopterygian fishes: Nephrotus and Calobodus, with remarks on the dentition of other forms. Zool Scr 7:297-326

Otto H (1908) Die Beschuppung der Brevilinguier und Ascalaboten. Jener Z Naturwiss 44:193-252

Owen R (1840-45) Odontography, or a treatise on the comparative anatomy of the teeth. Balliere, London

Paine G (1937) Fossilisation of bone. Am J Sci 234:148-157

Pales L (1930) Paléopathologie et pathologie comparative. Masson, Paris

Peabody FE (1961) Annual growth zones in vertebrates (living and fossil). J Morphol 108:11-62

Perrin WF, Mirrick AC (1980) Age determination of toothed whales and sirenians. Int Whaling Commiss Rep IWC Spec Issue, Cambridge

Petersen H (1930) Die Organe des Skelettsystems. In: Mollendorf V (ed) Handbuch der mikroskopischen Anatomie des Menschen. Springer, Berlin pp 529-676

Peyer B (1931) Hartgebilde der Integuments. In: Bolk L, Goppert E, Kallius E, Lubosch W (eds) Handbuch Vergleichende Anatomie der Wirbeltiere, vol 1. Urban & Schwarzenberg, Berlin, pp 703-752

Peyer B (1968) Comparative odontology. Univ Press, Chicago.

Queckett JT (1849a) On the intimate structure of bone as composing the skeleton in the four great classes of animals, etc. Trans Microsc Soc London 2:46-58

Queckett JT (1849b) Additional observations on the intimate structure of bone. Trans Microsc Soc London 2:40-42

Queckett JT (1855) Descriptive and illustrated catalogue of the histological series contained in the Museum of the Royal College of Surgeons of England, vol 2. London

Recker RR (ed) (1983) Bone histomorphometry: techniques and interpretation. CRC, Boca Raton

Reid REH (1981) Lamellar-zonal bone with zones and annuli in the pelvis of a sauropod dinosaur. Nature (London) 292:49-51

Reid REH (1983) High vascularity in bones of dinosaurs, mammals and birds. Geol Mag 120:191-194

Reid REH (1984) The histology of dinosaurian bone, and its possible bearing on dinosaurian physiology. In: Fergusson MWJ (ed) The structure, development and evolution of reptiles. Symp Zool Soc London 52:629-663

Reid REH (1985) On supposed Haversian bone from the hadrosaur Anatosaurus and the nature of compact bone in dinosaurs. J Paleontol 59:140-148

Reif WE (1982) Evolution of dermal skeleton and dentition in vertebrates - the odontode regulation theory. Evol Biol 15:287-368

Rhodin AGJ (1985) Comparative chondro-osseous development and growth of marine turtles. Copeia 3:752-771

Ricqlès A de (1974) Evolution of endothermy: histological evidences. Evol Theor 1:51-80

Ricqlès A de (1975a) Quelques remarques paléohistolgiques sur le problème de la néotenie chez les stégocéphales. In: Lehman JP (ed) Problèmes actuels de palèontologie. evolution des vertébrés, vol 1. CNRS, Coll 218, Paris, pp 351-363

Ricqlès A de (1975b) On bone histology of fossil and living reptiles, with comments on its functional and evolutionary significance. In: Cox B, Bellairs A (eds) Morphology and biology of reptiles. Linn Soc Symp 3:123-150

Ricqlès A de (1975-78) Recherches paléohistologiques sur les os longs des tétrapods VII. Ann Paléont 61:51-129; 62:71-126; 63:33-56; 63:133-160; 64:85-111; 64:153-184

Ricqlès A de (1979) Quelques remarques sur l'histoire évolutive des tissus squelettiques chez les vertébrés et plus particulièrement chez les tétrapodes. Ann Biol 18:1-35

Ricqlès A de (1980) Tissue structures of dinosaur bone. Functional significance and possible relation to dinosaur physiology. In: Thomas RDK, Olson EC (eds) A cold look at warm blooded dinosaurs. AAAS Symp 28, Boulder, pp 103-139

Ricqlès A de (1986) On the usefulness and prospects of bone histology for vertebrate paleontology. Paleopath Assoc News 5:16

Ricqlès A de (1989) Les mécanismes hétérochroniques dans le retour des tétrapodes au milieu aquatique. In: Chaline J (ed) Ontogenèse et évolution. Geobios Mem Spec, Lyon, pp 337-348

Ricqlès A de, Bolt JR (1983) Jaw growth and tooth replacement in Captorhinus aguti (Reptilia: Captorhinomorpha): a morphological and histological analysis. J Vertebr Paleontol 3:7-24

Ricqlès A de, Meunier FJ, Castanet J, Francillon-Vieillot H (in press) Comparative microstructure of bone. In: Hall BK (ed) Bone matrix and bone specific products, vol 3

Rogers AF (1924) Minerology and petrolography of fossil bone. Bull Geol Soc Am 35:535-556

Rotschild BM (1987a) Diffuse idiopathic skeletal hyperostosis as reflected in the paleontological record: dinosaurs and early mammals. Seminars in Arthritis and Rheumatism 17:119-125

Rotschild BM (1987b) Decompression syndrome in fossil marine turtles. Ann Carnegie Mus 56:253-258

Rotschild BM (1988) Stress fracture in a ceratopsian phalanx. J Paleontol 62:302-303

Roumiantsiev AV (1958) Essay on the evolution of cartilaginous and bony tissues. Akad Nauk SSSR, Moscow (in Russian)

Schaeffer B (1977) The dermal skeleton in fishes. In: Andrews SM, Miles RS, Walker AD (eds) Problems in vertebrate evolution. Linn Soc, London, pp 25-52

Schaffer J (1889) Über den feineren Bau fossiler Knochen. Sitzungsber Kaiserl Akad Wiss Math Naturwiss Kl 3-7:319-382

Schmidt WJ, Kiel A (1971) Polarizing microscopy of dental tissues. Pergamon, Oxford

Schultze HP (1966) Morphologische und histologische Untersuchungen an Schuppen mesozoischer Actinopterygier. Übergang von Ganoid-zu Rund-Schuppen. N Jahrb Geol Palaontol 126:232-314

Schultze HP (1969) Die Faltenzähne der Rhipidistiiden Crossopterygier, der Tetrapoden und der Actinopterygier-Gattung Lepisosteus. Palaeontol It 65:63-136

Schultze HP (1977) Ausgangsformen und Entwicklung der rhombischen Schuppen des Osteichthyes (Pisces). Palaontol Z 51:152-168

Seitz AL (1907) Vergleichende Studien über den mikroskopischen Knochenbau fossiler und rezenter Reptilien. Nova Acta Akad Leop Carol Naturforsch 87:230-370

Shipman P (1981) Life history of a fossil. Harvard Univ Press, Cambridge

Smith JW (1960) Collagen fibre patterns in mammalian bone. J Anat 94:329-344

Smith MM (1988) The dentition of paleozoic lungfishes: a consideration of the significance of teeth, denticles and tooth plates for dipnoan phylogeny. Mem Mus Nat Hist Nat 53:179-194

Stephan P (1900) Recherches histologiques sur la structure du tissu osseux des poissons. Bull Soc Fr Belg 33:281-429

Tchudinov PK (1970) About skin structures among therapsids (Estemmenosuchus uralensis). Nauka, Moscow, pp 45-50

Thomasset JJ (1930) Recherches sur les tissus dentaires des poissons fossiles. Arch Anat Histol Embryol 11:5-153

Thomson KS (1975) On the biology of cosmine. Bull Peabody Mus Nat Hist 40:1-59

Tomes CS (1878) On the structure and development of vascular dentine. Phil Trans R Soc London 169:1-3

Tomes CS (1898) Upon the structure and development of the enamel of the elasmobranch fishes. Philos Trans R Soc London 190:7

Tretjakoff D & Chinkus F (1927) Das Knochengewebe der Fische. Z Anat Entwickl 83:363-396

Tretjakoff D & Chinkus F (1930) Die Hautschuppen der Anamnia. Z Wiss Zool 136:3-4

Valentin (1861) Die Untersuchungen der Pflanzen- und Tiergewebe im polarisierten Licht. Leipzig

Waddington CH (1957) The strategy of the genes. Alan & Unwin, London

Walcott CD (1892) Preliminary notes on the discovery of a vertebrate fauna in silurian (ordovician) strata. Bull Geol Soc Am 3:153-172

Weidenreich F (1930) Das Knochengewebe. In: Mollendorff V (ed) Handbuch der mikroskopischen Anatomie des Menschen, vol II,2. Springer, Berlin, pp 391-520

Williamson WC (1849) On the microscopic structure of the scale and dermal teeth of some ganoid and placoid fish. Philos Trans R Soc London 139:435-475

Williamson WC (1851) Investigations into the structure and development of the scales and bones of fishes. Philos Trans Roy Soc London 141:643-702

Wurmbach H (1932) Das Wachstum des Selachierwirbels und seiner Gewebe. Zool Jahrb 55:1-136

Zylberberg L, Wake MH (1990) Structure of the scales of Dermophis and Microcaecilia (Amphibia: Gymnophiona), and a comparison to dermal ossifications of other vertebrates. J Morphol 206:25-43

Differential Diagnosis of Human and Animal Bone

Laszlo Harsányi
Department of Forensic Medicine
University Medical School of Pécs
Sziget 12
H-7643 Pécs
Hungary

The first step in the anthropological and forensic examination of skeletal fragments recovered from archaeological excavations or elsewhere is to decide whether the finds are of human origin. This question can be decided by employing the following procedures:

1. Comparative anatomical, macroscopic examination
2. Analysing X-rays
3. Precipitation, immunological reaction
4. Immunoelectrophoresis
5. Fluorescent immunohistological procedures
6. Light microscope examination
7. Scanning electron microscope examination

1 Anatomical Examination

Human origin is easy to prove or exclude in large bones or in fragments of adult bones which demonstrate characteristic adult features. Nevertheless, certain bones originating from animals, especially those from small animals, may give grounds to suspect human origin. The false judgement in most of these cases is based on mistaking animal bones for those of neonates or infants. The origin of carpal bones and those of the foot may also be difficult to decide, however, sufficient knowledge of anatomy frequently helps to avoid misjudgment.

Parisot and Mutel (1929) were able to determine the human origin of one single distal phalanx, excluding its origin from an ape, by relying on the tuberositas unguicularis being specifically wide and flat in humans.

2 Analysing X-Rays

X-rays of long bones often demonstrate species-specific structural differences. The intactness of the whole bone or the majority of the diaphysis is indispensable for correct evaluation of the radiograph (Depreux and Muller 1953; Chilvarquer el al. 1987). An advantage of this procedure is that it does not damage the find. The reliability of the X-ray evaluation was found to be approximately 75%. On the basis of 20 cases examined radiologically by Chilvarquer et al. (1987), they obtained the correct species diagnosis in 81.9-86.8% of specimens.

3 Precipitation Reaction

In 1899 Tchistovich, a Russian researcher working in the Pasteur Institute in Paris, found that gel proteins induce the formation of species specific precipitates in various other animal species. Uhlenhuth, unaware of the Russian researcher's work, arrived at the same conclusion in 1901. He recommended the procedure as one to be used in order to determine the human or animal origin of blood stains. Beumer (1914) used the Uhlenhuth method for determining the species to which bones belonged. He made a physiological salt solution extract from 0.5 g of fine-ground compact bone, and then performed the precipitating reaction. This immunological technique is more sensitive if it is used not in a liquid medium but in gelatin agar; see Ouchterlony (1953). However, 40 years after burial, species specific proteins decompose to such an extent that the reaction becomes negative (Berg and Specht 1958). This is why, in human remains, negative findings in precipitate reactions definitely indicate that the age of origin is more than 40 years. This conclusion is of great importance in forensic medicine. Negative species specific serological reactions can also be expected when the bones are exposed to heat of not less 130-150 °C (Steffenhagen and Clough 1910).

4 Immunoelectrophoresis

Immunoelectrophoretic examination of an extract of compact bone buried for not more than 40 years can be performed with polyvalent antihuman serum, as a result of which the various protein fractions can also be differentiated.

5 Fluorescent Immunohistology

The fluorescent immunohistological method is appealing, easy to demonstrate and not too difficult to perform. By using various kinds of tissues it is applicable in blood group determination (for details see, for example, Coons and Kaplan 1950; Glynn and Holborow 1959). The human origin of bone fragments buried for not more than 40-50 years can be determined by employing immune serum marked with fluorescent isothiocyanate (Harsányi 1976a, 1978). Unfortunately, the three latter techniques cannot be used for archaeological finds.

6 Light Microscope Examination

In 1903 Kenyeres and Hegyi, two Hungarian researchers, reported that cross sections of compact bone of the diaphyses of long bones can be used to determine the human origin of bone fragments, as the average diameter of the Haversian canals in human bones is significantly greater than that of animal bone tissues; an observation which has been confirmed by others. Hey (1924) remarked that various pathological processes, for example sclerosing osteitis, rickets, osteomalacia, inflammation, and tumours, may result in structural changes of the bony tissue and therefore, their histological examination may give misleading results. Histological differences between bony tissues of human and animal origin are present during the whole life span, thus it is impossible to confuse the various species or become confused in the determination of the human origin of bones of neonates, infants or persons having suffered from senile marasmus. The basic structural types of bony tissues are determined by the vascularization of the bones. Hinüber (1951), Förster and Goldbach (1954) and Goldbach and Hinüber (1955) have presented practical descriptions based on the study of 528 long bones of humans and various mammals (see Table 1).

Considering structure, the compact substance of human long bones belong to the non-layered and longitudinally vascularized type of bones, which are composed of solitary osteons, round and oval in shape and Haversian canals which are mostly excentric in position.

Transverse slabs of a thickness of 5-6 mm of certain sectors may be removed from the compact substance of long bone and prepared for histological examination. The slabs are fixed in 10% neutral formalin and are decalcified. Decalcification can be performed using an acid decalcifying agent; however, an aqueous solution of ethylenediaminetetraacetic acid (EDTA) has been shown to do less harm to

Table 1 Structure of the compact substance
(After Goldbach and Hinüber 1955)

A. General structure:
   1. Non-laminated
   2. Multi-laminated
   3. Multi-laminated, compact

B. Vascularization:
   1. Mainly longitudinal
   2. Mainly circular
   3. Mainly radial

   Cross section of vessels:
   1. Regular
   2. Partially regular
   3. Irregular

C. Osteon types:
   1. Linear osteons
   2. System of solitary osteons
   3. Gyrose osteons
   4. Clusters of osteons
   5. Osteon lines

---

the structure. The process proceeds slowly, and decalcification may take as long as 1 or 2 weeks. The remaining organic substance can be studied in embedded or frozen sections. The slabs can be embedded in celloidin or paraffin. With either method of embedding or freezing, the section should be cut to a thickness of 15-20 μm. Making comparatively thick sections is highly advisable with archaeological finds, as, due to the slow erosion of the structure, sections thinner than that may be more difficult to evaluate. It is also advisable to determine the diameter of as many as 50-100 Haversian canals with a micrometer and then calculate the arithmetic mean of the canals. The canals are comparatively narrow around the periosteal surface and wider near the medullary cavity. Therefore, measurements should

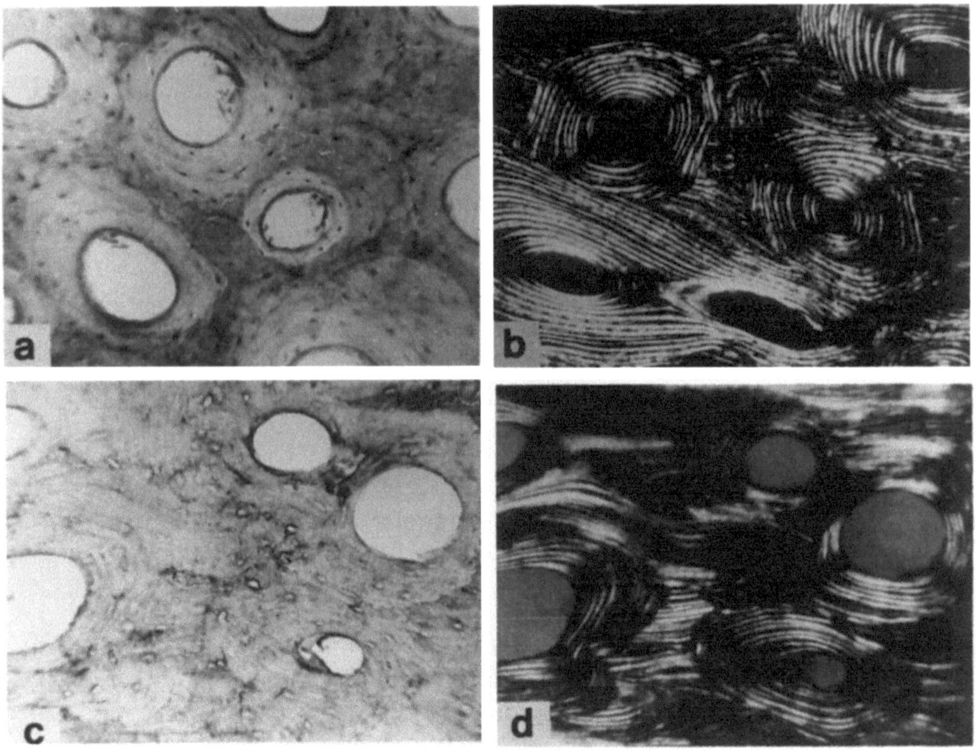
a
b
c
d

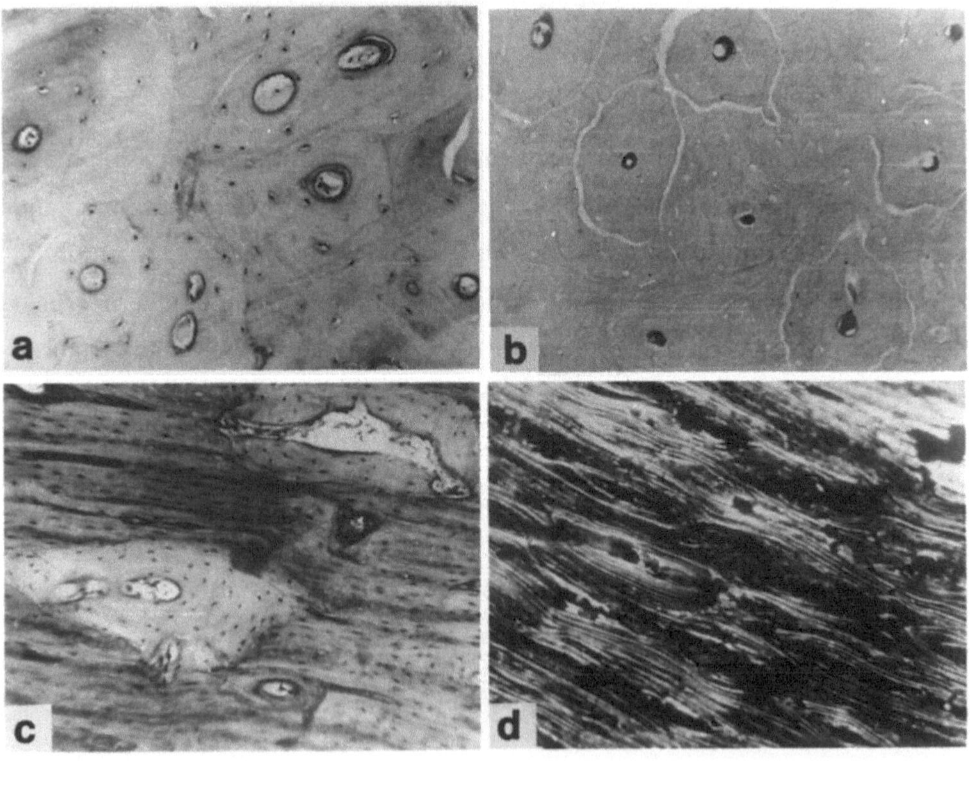
a
b
c
d

be started from the periosteum and continued step by step inwards, thus ensuring that the mean value includes the sizes of both the narrow and wide canals. When measuring the diameter of elliptical, slantwise canals, the lesser diameter should be taken into account (Figs. 1 and 2).

The author of the present study has unfortunately been unable to study specimens which have not previously been decalcified owing to a lack of the appropriate device.

In their studies Rämsch and Zerndt (1963), Gladühsev (1964) and others confirmed the correctness of the observations stated above. The histological structure was classified according to the diameter of the canals. They established the following classification of diameters:

Classification	Diameter
Very short	--> 10 μm
Short	11 - 20 μm
Medium long	21 - 40 μm
Long	41 - 80 μm
Very long	80 μm -->

Diameters of less than 40 μm in length are rarely found in human bones, the average diameter being approximately 60 μm. The classification also includes the recognizable number of canals/visual field, however, this number depends on the degree of magnification and, thus, it cannot be considered as a basic value (Table 2).

Fig. 1 Humerus from recent human remains
- **a** Stain: H & E magnification x265;
- **b** Birefringence by polarized light. Stain: H & E magnif. x265;
- **c** Human 10th century, humerus. Stain: H & E magnif. x265;
- **d** Birefringence by polarized light. Stain: H & E magnif. x265

Fig. 2
- **a** Horse;
- **b** Cattle;
- **c** Sheep;
- **d** Sheep bone under polarized light. Stain: H & E magnif. x265

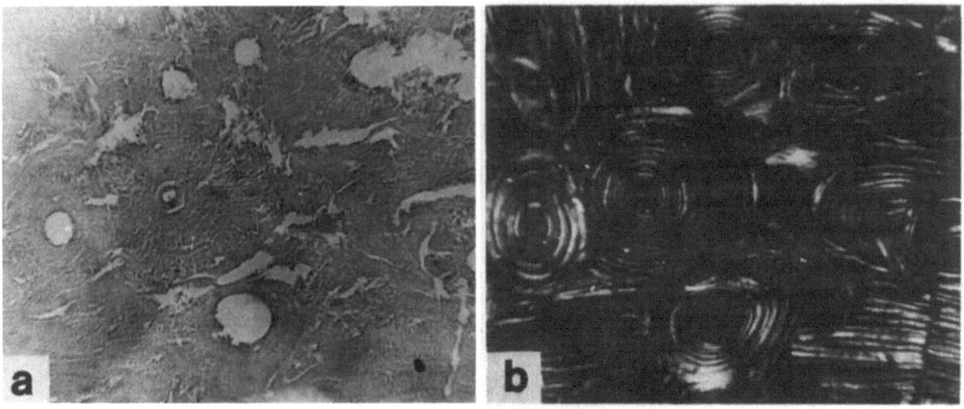

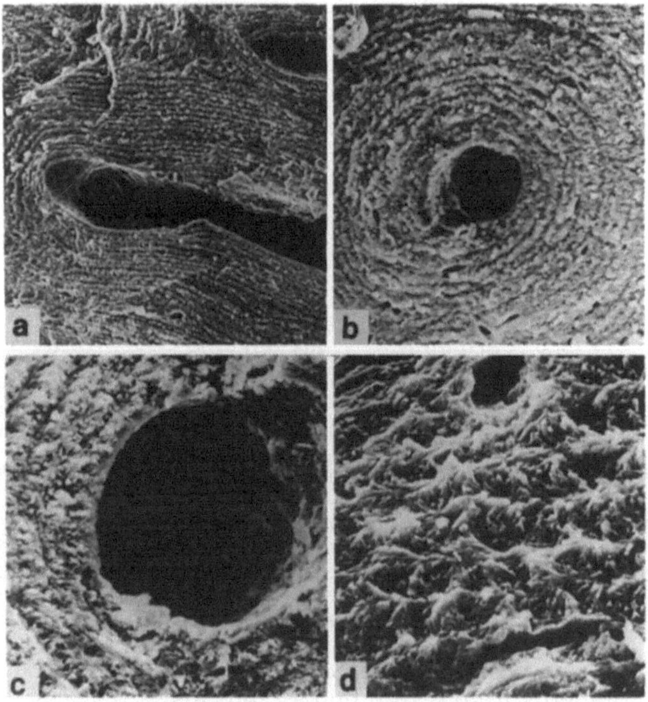

Fig. 3 Cave bear, approximately 25000 years b.c.
a Stain: H & E;
b Section seen under polarizing light; magnification x265

Fig. 4 Scanning electron microscope images of human bone cremated at approx. 550-600°C.
a Magnification x300;
b Magnification x500;
c Magnification x1500;
d Magnification x3000

Table 2 The diameter and number of Haversian canals (after Müller and Demarez 1934)

Species	Average diameter µm	Average number of Haversian canals per visual field	Structural features
Human, neonate	54.3	2.3	
Human, 6 mo.	60.5	1.7	
Human, 12 mo.	71.6	1.6	several osteons with long diameter oval an round medium and very large
Human, 18 mo.	56.8	1.7	
Human, 41 y	52.9	1.7	
Human, 70 y		1.5	
Horse	30.0	2.7	from very small to large, mostly small
Cow	47.9	1.4	medium sized, irregular shapes
Goat	21.2	2.4	mostly medium smaller near medullary cavity
Sheep	18.2	3.6	medium, irregular
Pig	32.8	2.1	mostly medium, smaller near medullary cavity
Dog	21.2	3.0	mostly very small
Rabbit	12.6	8.0	very small
Cat	20.3	2.8	mostly very small
Hen	14.0	7.0	very small, round
Goose	15.7	14.1	small
Ape	30-40		

The birefringence of the bone collagen in polarized light can be studied in both recent and archaeological specimens. The decrease in intensity of birefringence shows direct correlation with the increase in time of origin. When the quantity of bone collagen shows significant decrease and falls below 1 g% of the total nitrogen content of the solid material, the intensity of birefringence becomes remarkably low (Fig. 3). The tubular structure of bones found in various species of animals of the same size was investigated by the author of the present study, however, no clear distinction could be made, for example, between the solid material of goat bones and that of lamb bones. Neither were the characteristic structural features of flat bones suitable for setting clear diagnoses. However, the experience gained in the fields of criminology and forensic medicine and cooperation with anthropologists gave sufficient grounds for determining human origin. Neither Rämsch and Zerndt (1963) nor the present author had the opportunity to study long bones of wild animals and to compare them with those of domestic animals, though this would be important, especially a study of the bone structure of apes.

7. Scanning Electron Microscope Examination

Regular study of substances after incineration or cremation of cadavers offers much useful data. Cremation of the dead with burial of the remains in urns is a tradition which arose in Europe and in other parts of the world. The possibilities and limits of studying the finds is dealt with in several reports (see, for example, Janssens 1970). When cadavers are incinerated and the bones become heated, after denaturation of proteins, the water content of bone is removed between 300°C and 500°C and the CO_2 content at about 600°C. Above 700°C the apatite crystals lose their water of crystallization but, on cooling again, they become recrystallized and regain the apatite shape. These "secondary" crystals, however, may well be 50 times as large as the original ones. The apatite shape is stable so long as the temperature does not exceed 1200°C (Burri et al. 1935; Harsányi 1976, 1977). The calcined and incinerated bone remains cannot be studied by the usual histological methods because of the fragility of the finds.

These finds are suitable for scanning electron microscope studies. Herrmann (1972) was the first to employ this method in studying finds of anthropological importance. "Homo aurignaciensis hauseri", a *Homo sapiens* find discovered in France in 1909, was sold to the Museum für Volkerkunde, Berlin in 1910. As a consequence of a bomb attack followed by a fire in 1945 the invaluable find was almost completely incinerated. Herrmann identified the burnt remains found among the ruins and, using several methods including scanning electron microscopic examinations, studied the damage caused by the extreme heat effect.

Cadavers are incinerated at comparatively low temperatures. Berg et al. (1981) found that at 500°C cadavers of neonates and infants are fully incinerated and only very few bone remains can survive. The quantity of substance left after cremation is very low. The weight of ashes found after cremating cadavers in modern electric furnaces was 1800-2000 g for males and 1500-1700 g for females (Malinowski and Porawski 1969; Herrmann 1976). Finds of substance resulting from cremation, to be studied in the course of anthropological examinations constitute only a part of the original quantity of ashes, as it was obviously impossible to collect all the remains of the cadaver after cremation. On the other hand, ashes of cadavers and the fuel, in most cases wood, are found together; Nemeskéri & Harsányi (1968) studied the skeletal remains of 17 middle Bronze Age persons buried following cremation and found the total weight of material to be only

5220 g, of which the average weight of remains (human) was 307 g. Certain parts of the diaphyses of long bones can in most cases be separated from the small and fragmentary remains and, with some difficulty, transverse fractures or sawed surfaces can be made, which are suitable for scanning electron microscopice examination. In order to measure the diameter of at least 30-50 Haversian canals, several specimens must be examined. The fact that the volume of bony tissues exposed to the effects of extreme heat decreases must be taken into account. The degree of "shrinkage" is significantly high longitudinally. Bones of neonates and infants contract to a greater extent; the degree of contraction averages 10%, which, when evaluating material, must be taken into consideration. The usual temperature used for cremation is 500-700 °C. Thus, such materials were not damaged by the high temperature electric furnaces are operated at. Thus the inorganic substance did not "flow away", consequently it did not turn into a kind of solid, glasslike, almost white and porcelain-clinking material, which lost its original microstructure (Fig. 4).

In experiments the author of the present study used scanning electron microscope examination to study the morphological changes in diaphyses of long bones which had been heated to between 300 and 1300 °C. Characteristic features of the periosteal and fracture surfaces were observed. Water, which constitutes 15-20% of bone, is removed at 300 °C over the course of 30-60 min and the lamina fundamentalis externa lifts and forms "blisters" whose walls burst. At the same time burst blisters appear on the outer surface. The original structure does not change as long as the temperature the bone is exposed to is below 700°C. The osteons may become separated from one another in the vicinity of the periosteal surface, however, they preserve their structural unity. Temperatures above 700°C damage the compact substance to such an extent that the Haversian canals, the osteons and their original diameters become impossible to evaluate. Specimens, as a rule, were coated with gold in a vacuum of 4×10^{-5} Torr. No graphite was employed under the layer of gold.

8. Conclusions

1. Light microscope examination of cross sections made of diaphyses of long bones can be used for the determination of human origin by measuring the average diameter of the Haversian canals.
2. The diameter of at least 50-100 Haversian canals should be measured and those in the vicinity of both the periosteal surface and the medullar cavity must be taken into account.
3. Diameters of the Haversian canals found in human long bones are significantly longer lasting from infancy until senile atrophy than those found in bones of animal origin.
4. If signs of tissue diseases, inflammation or tumours are found in the material under study, the evaluation of the histological structure is not feasible.
5. In domestic animals of the same size the various species cannot be distinguished from one another by studying the structure of their bony tissue.
6. The author has not had the opportunity to study the bony tissue of wild animals, particularly apes; this would be an important study.
7. When studying bone remains originating from incinerated or cremated cadavers by scanning electron microscope examination, the decrease in bone volume due to the exposure to extreme heat must be taken into account. The degree of this decrease is higher in bones of neonates or infants, compared to those in adults where it never exceeds on average

13.55% (Herrmann 1977).

References

Berg S, Specht W (1958) Untersuchungen zur Bestimmung der Liegezeit von Skeletteilen. Dtsch Z Gerichtsmed 47:209-241

Berg S, Rolle R, Seemann H (1981) Der Archäologe und der Tod. Bucher, München Luzern

Beumer R (1914) Die Unterscheidung von Menschen- und Tierknochen. In: Lochte T (ed) Gerichtsärztliche und polizeiärztliche Technik. Bergmann, Wiesbaden, pp 286-290

Burri C, Jakob J, Parker RL, Strunz H (1935) Über Hydroxylapatit von der Kemmleten bei Hospenthal. Schweiz Mineral Petrogr Mitt 15:327-339

Chilvarquer I, Katz JO, Glassman DM, Prihoda TJ, Cottone JA (1987) Comparative radiographic study of human and animal long bone patterns. J Forens Sci 32:1645-1654

Coons AH, Kaplan MH (1950) Localization of antigen in tissue cells. J Exp Med 91:1-13

Depreux R, Muller P (1953) Différenciation radiologique des os humains et animaux. IN: Trav 26 Congr Int de Med Leg, pp 204-212

Förster A, Goldbach H-J (1954) Die histologische Differnzierung von Femurdiaphysen Neugeborener, Kleinkinder und kleiner Haustiere. Dtsch Z Gerichtsmed 43:273-289

Gladüshew JM (1964) Problems of the histological investigation of the bone in forensic medicine. Sudebnomed Exp 7:fasc 2 23-26 (in Russian)

Glynn LE, Holborow EJ (1959) Distribution of blood group substances in human tissues. Brit Med Bull 15:150-153

Goldbach HJ, Hinüber H (1955) Versuch einer Systematik der Formelemente des Säugetierknochens. Dtsch Z Gerichtsmed 44:578-588

Harsányi L (1976a) Determination of species in bone tissue by means of fluorescent antibody. Morphol Ig Orv Szle 16:100-102 (in Hungarian)

Harsányi L (1976b) Scanning electron microscopic investigation of thermal damage of the teeth. Acta Morphol Acad Sci Hung 23:271-281

Harsányi L (1977) Scanning electron microscopic investigation of thermal damage of the bone tissue. Morpol Ig Orv Szle 17:26-32 (in Hungarian)

Harsányi L (1978) Unterscheidung von Menschen- und Tierknochen. In: Hunger H, Leopold D (eds) Identifikation. Springer, Berlin Heidelberg New York, pp 100-112

Herrmann B (1972) Das Combe Capelle-Skelett. Ausgrabungen Berlin 3:7-69

Herrmann B (1976) Neuere Ergebnisse zur Beurteilung menschlicher Brandknochen. Z Rechtsmed 77:191-200

Herrmann B (1977) Über die Abhängigkeit der Schrumpfung vom Mineralgehalt bei experimentell verbrannten Knochen. Anthropol Anz 36:7-12

Hey (1924) Die histologische Identifizierung von menschlichen und tierischen Knochen. Dtsch Z Gerichtsmed 4:566-576

Hinüber H (1951) Die Unterschiede im Feinbau von Menschen- und Tierknochen als gerichtsärztliches Differentialdiagnostikum. Ph.D. Thesis, Univ Marburg

Janssens PA (1970) Palaeopathology diseases and injuries of prehistoric man. Baker, London

Kenyeres B, Hegyi M (1903) Unterscheidung des menschlichen und tierischen Knochengewebes. Vierteljahresschr Gerichtsmed 26:225-232

Malinowski A, Porawski R (1969) Identifikationsmöglichkeiten menschlicher Brandknochen mit besonderer Berücksichtigung ihres Gewichts. Zacchia 44:392-410

Muller M, Demarez R (1934) Le diagnostic différentiel de l'os de singe et de l'os humain. Med Leg 14:598-560

Nemeskéri J, Harsányi L (1968) Problems of the investigation of cremated bones. Anthropol Közl 12:99-116 (in Hungarian)

Ouchterlony Ö (1953) Antigen-antibody reactions in gels. Acta Pathol Microbiol Scand 32:231-244

Parisot P, Mutel M (1929) L'importance de la phalange unguéale dans l'identification de débris osseux. Ann Med Leg 9:543-545

Rämsch R, Zerndt B (1963) Vergleichende Untersuchungen der Haversschen Kanäle zwischen Menschen und Haustieren. Arch Kriminol 131:74-87

Steffenhagen K, Clough PW (1910) Biologische Untersuchungen über die Herkunft von Knochen. Berliner Klin Wochenschr 46:2097-2102

Tchistovitch Th (1899) Sur l'immunisation contre le sérum d'an guilles. Ann Inst Pasteur 13:406-425

Uhlenhuth P (1901) Eine Methode zur Unterscheidung der verschiedenen Blutarten, im besonderen zum differentialdiagnostischen Nachweis des Menschenblutes. Dtsch Med Wochenschr 26:82-83

Human Bone Remodelling and Aging

Hilde Uytterschaut
Laboratorium voor Anatomie en Embryologie
University of Groningen
Oostersingel 69
NL-9713 EZ Groningen
The Netherlands

1 Introduction

In 1965 Kerley developed a histological method for the determination of age of human bones (Kerley 1965), based on the transformation of bone during the ageing process; that is, the turnover of primary bone into secondary bone. In 1969, Ahlqvist and Damsten introduced some changes and improvements to this method (Ahlqvist and Damsten 1969).

However, despite several important advantages of these microscopic methods, as compared to age determination by means of macroscopic methods, application of the former methods is still relatively scarce. There are several reasons for this; firstly, the technical problem of preparing thin sections of bone; secondly, effects of field size when using microscopic methods (Ohlah 1977; Kerley and Ubelaker 1978; Stout and Gehlert 1982); and thirdly, difficulties in distinguishing the four structures differentiated by Kerley (Ahlqvist and Damsten, 1969).

2 Material and methods

The material used in this study included 20 femora and 20 tibiae from 20 individuals which were obtained from the Department of

Anatomy dissecting room in Leiden. The sex and age of the individuals were known with certainty. Because no significant sex differences were noted (Kerley 1965; Singh and Gunnberg, 1970) no attention was paid to the sex and the sample of 20 individuals was built up according to their age. The 20 individuals were 17-92 years old (mean age was 54 years) and showed about the same distribution of ages as in the study by Ahlqvist and Damsten (mean age of their sample was 55 years). It was not that easy to build up a sample of 20 femora and 20 tibiae from 20 individuals between ±20 and ±90 years old. Many of the bones obtained from the dissecting room turned out to have a damaged cortex, which made application of histological ageing methods impossible.

Kerley (1965) and Ahlqvist and Damsten (1969), but also Frost (1958), Singh and Gunberg (1970), Ubelaker (1974) and Stout (1976), made use of ground sections. Unfortunately, this technique has some inherent technical problems: firstly, the speed at which the saw grinds through the bone is not easy to adjust; secondly, by mounting the bone on a slide, the expansion of the bone may crack the glass slide (Ubelaker 1974); and thirdly, the ground sections may not be sufficiently thin for detailed histological analysis, and are often distorted by the grinding process (Stout and Teitelbaum 1976).

Advances in technology have led to the development of different kinds of microtomes, e.g. saw microtomes. One of the advantages of saw microtomes over other microtomes, e.g. sledge microtomes (Stout and Teitelbaum 1976), is the elimination of the microscopically observable "planing-ice" effect in the sections so that the anatomical structures in the sections remain intact. Saw microtomes (Fig. 1) comprise of a turning diamond-disc which saws through the bone at a speed adjustable by the user, depending on the desired thickness of the section. For cooling purposes the disc is moistened during sawing.

The midshaft part (within 1-2 inches of the middle of the shaft) of the bones was sawn with a saw microtome into sections of ±30 μm. The bone material was not decalcified. Initially the bone was embedded in methyl-methacrylate, but later it turned out that this was not really necessary as each of the 40 bones could be sawn without being embedded. Only for more fossilized and brittle material was embedding of the bone necessary.

The unstained sections were studied by light microscopy using polarized light, a method which was also used by Kerley and even as early as 1878 by Aeby and Schaffer (Schaffer 1889) for the study of fossil bones (Stout and Gehlert 1979).

Although, according to Stout and Gehlert (1980), Kerley's method seemed to be more accurate and reliable, I would advocate the method of Ahlqvist and Damsten as the method of choice based not only on practical arguments but also with regard to content.

Ahlqvist and Damsten differentiated between two categories: (intact) osteons and osteon fragments (first category) and lamellar bone and non-Haversian canals (second category). Kerley made a distinction between each of the four above mentioned structures, but, unfortunately, unambiguous identification of one or more of the four structures is not always possible (Ubelaker 1974). Moreover, there are no precise definitions for some of the parameters, e.g. Stout and Gehlert (1980) found four different definitions for "intact osteon" in applications of histological ageing methods. Another disadvantage of Kerley's method lies in the anatomical field location: one of the four fields in the cross-section falls upon the linea aspera, a location where histological structures are less well correlated with age, possibly because of the muscle insertions on the femoral crest (Ahlqvist and Damsten 1969). Bouvier and Ubelaker (1977) found that it was difficult to obtain a reliable count from this region. A few remarks regarding studies which make comparisons between the accuracy of both methods will be mentioned in the following discussion. These remarks led to the conclusion that Ahlqvist and Damsten's method, when compared to other methods, is an accurate method not only for younger but also for older specimens.

Beside these aspects of content, there are some practical advantages of employing Ahlqvist and Damsten's method. In Kerley's method the number of different structures is counted within a circular visual field. Ahlqvist and Damsten, on the other hand, used a square-ruled network containing 100 squares superimposed on the sections; the number of squares more than half filled with osteon and osteon fragments were counted. As Ubelaker (1969) pointed out, it is easier to count squares in the reticule than to count actual structures.

With respect to the field size, the method introduced by Kerley raised some confusion, difficulties and changes in its application. In 1978 Kerley and Ubelaker published a revision of the regression formulae. This revision was necessary because it was found that the field size used in Kerley's original study was not 1.25 mm but 1.62 mm. Because of the large range in field sizes produced by different microscopes, 10x objectives and 10x oculars (Stout and Gehlert, 1982), Kerley and Ubelaker suggested investigators should adjust the data to a field diameter of 1.62 mm. The area of the 100x field of the microscope that is used must be calculated using a stage micrometer and this field size must be divided by 2.06 mm^2 (area of a field diameter of 1.62 mm). All counts of osteons, fragments or non-Haversian canals should then be multiplied by this factor. However, in 1982, Stout and Gehlert pointed out that correction factors cannot equalize the counts of osteons and osteon fragments because of spatial variations in the distributions of these histological structures.

In practice, this study proceeded as follows. The Ahlqvist and Damsten method was applied by using a microscope with a drawing attachment and with a 10x objective and 10x ocular. An area of

1 mm² was measured with a stage micrometer and plotted on a sheet of paper. The resulting square was then divided into 100 squares. Ahlqvist and Damsten made use of an ocular square-ruled network and, in practice, it turned out that it was almost impossible to find a combination of a 10x objective with a 10x ocular and a network that produced a field size of exactly 1 mm². It is known that different microscopes, objectives, oculars and all possible combinations produce different-sized fields of view (Stout and Gehlert 1982; Kerley 1969).

3 Results

The percentage of osteons and osteon fragments was counted according to the principle of the histological method proposed by Ahlqvist and Damsten. This was undertaken once on each of the 20 femoral cross-sections and once on the 20 tibial cross-sections. The 20 femoral and 20 tibial cross-sections were derived from the same 20 individuals.

For the study of the femora the percentage of osteons and osteon fragments at four different anatomical fields of each section (as prescribed by Ahlqvist and Damsten 1969, p. 208) were estimated and their regression formula ($y = 0.991x - 4.96$, where y is the estimated age in years and x is the percentage of osteons and osteon fragments) was then applied. Furthermore, the regression line with best fit to the obtained data was constructed. Application of the regression formula suggested by Ahlqvist and Damsten led to a standard error of estimate of 6.51 years, and the mean difference between the actual ages and those estimated from the percentages of osteons and osteon fragments was revealed to be 3.38 years. Table 1, column 2, shows the percentages of osteons and osteon fragments for each of the 20 individuals, counted in the cross-sections of the femur. The correlation between the actual age and the percentage of osteons and osteon fragments was highly significant, and was found to be 0.959. These results led to the following regression formula: $y = 0.982x - 1.19$ with a standard error of 6.51 years. Figure 2 shows a plot of the regression line. The slopes and y-intercepts of this regression line are very good as compared with those of the regression line obtained by Ahlqvist and Damsten (1969). The slope of the regression line of Ahlqvist and Damsten was 0.991, the slope of the regression line obtained in this study was 0.982.

Having tested the method and the regression formula of Ahlqvist and Damsten to the femur, the applicability of this method was extended by constructing a regression formula for the tibia. Therefore, the midshaft part of the each tibia was sawn by a saw microtome into sections of $\pm$ 30 μm. The percentage of osteons and osteon fragments were then counted at six different fields of each section (Fig. 3): two fields centrally along the dorsal side of the bone, two fields centrally along the medial side and two fields centrally along the lateral side. For each side the two counts were averaged. The resulting means of each side were averaged again and the age at death was estimated. Two fields along each side of the section were examined in order to avoid "inadequate sampling", as described by Stout (1978). Table 1, column 3, presents the percentages of osteons and osteon fragments for the tibiae of the 20 individuals. Again, the correlation between the actual ages and these percentages turned out to be highly significant, viz. 0.961. These results led to the following regression formula: $y = 0.989x + 3.31$ with a standard error of 6.29 years. Figure 4 shows the plot of this regression line, the slope of which is 0.989.

Table 1 Percentages of osteons and osteon fragments and corresponding actual ages; sample size = 20

Actual age (y)	osteons and osteon fragments (x) (%) Femur	Tibia
17	21	23
19	20	15
22	21	17
28	36	24
38	41	36
38	41	40
45	54	48
48	58	45
50	54	46
52	38	37
54	63	47
59	58	60
62	57	58
65	73	76
68	79	72
73	79	67
78	75	69
81	81	73
85	81	85
92	88	81
y = 53.70	x = 55.90	x = 50.95

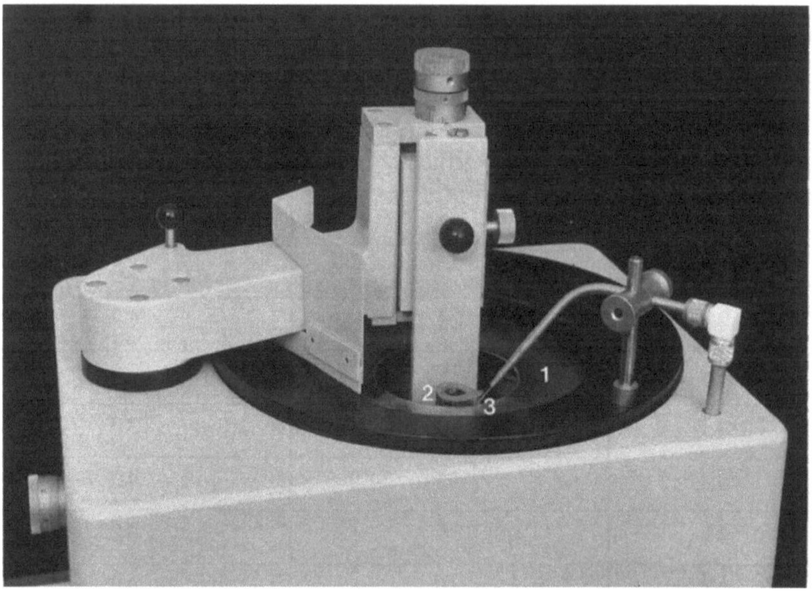

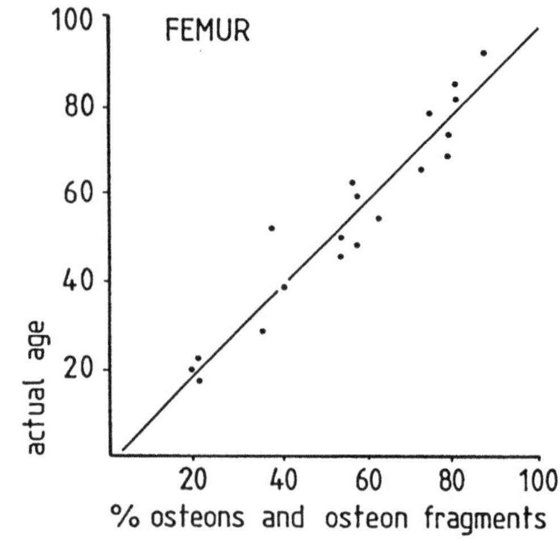

Fig 1 Saw microtome consisting of a diamond-disc (1) that saws through the bone (2). The disc is moistened by sprinkling (3) water onto it

Fig 2 Regression line showing the relationship between the percentage of osteons and osteon fragments in the femoral cross-section and the actual age

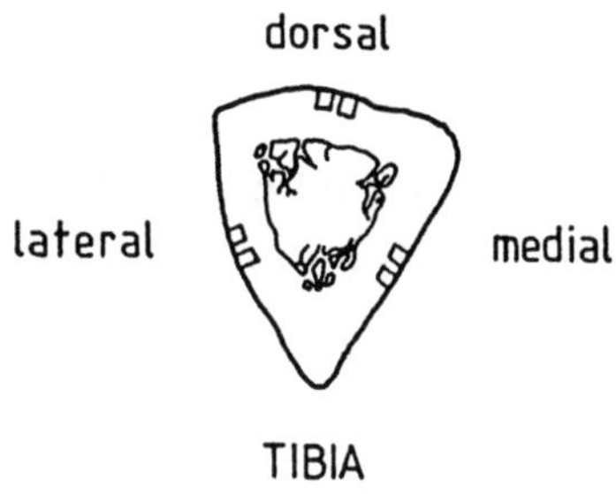

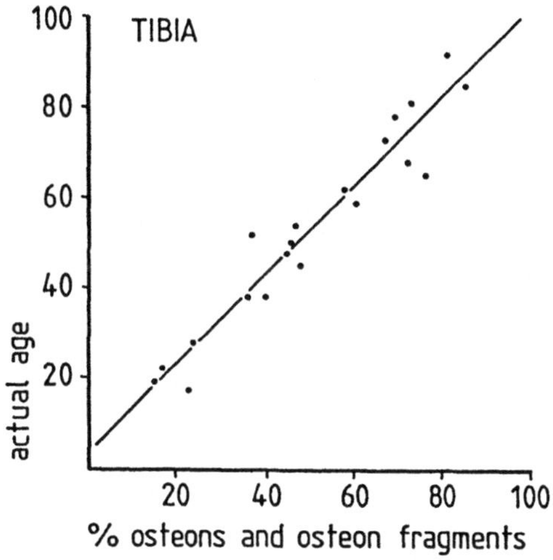

Fig 3 Cross-section of a tibia. The six square fields are the fields where the percentages of osteons and osteon fragments were counted

Fig 4 Regression line showing the relationship between the percentage of osteons and osteon fragments in the tibial cross-sections and the actual age

Ubelaker is one of the few people who has applied Kerley's and Ahlqvist and Damsten's method on large sample sizes. One of the studies in which he applied Kerley's method was the analysis of bone material excavated from Santo Domingo; the sample included 158 skeletons (Ubelaker 1981). The difference between the average of the actual ages of this sample (58.8 years) and the average of the estimated ages (54.6 years) was 4.2 years, a difference which is comparable to the difference of 3.38 years which were found in this study on the 20 femora. The standard error, however, was estimated to be 10.43 years, which is higher (significant at 0.05 level, one-tailed) than the standard error of 6.51 obtained here by the application of Ahlqvist and Damsten's method on the 20 femora.

As previously mentioned, some remarks with respect to the studies which make comparisons between the accuracy of Kerley's and Ahlqvist and Damsten's method are necessary. Firstly, for this kind of comparison it is necessary to use a new test sample and not the sample used in one of the original studies, as undertaken by Bouvier and Ubelaker (1977) (see Stout and Gehlert 1980). Secondly, the choice of the field size is problematical because Kerley used a field size of 2.06 mm^2 and Ahlqvist and Damsten used a field size of 1.00 mm^2. Although Stout and Gehlert in a later publication (Stout and Gehlert 1982) suggested use of a field size as close to 2.06 mm^2 as possible when employing Kerley's method, they still make use of the correction factors calculated by Kerley and Ubelaker (1978) in their comparison study (Stout and Gehlert, 1980). From the histograms made by Stout and Gehlert (1980) it can be seen that the accuracy of Ahlqvist and Damsten's method, but not of Kerley's, is more or less constant in both age groups (13-51 years and 60-102 years). Moreover, for the 60-102 years age range, the accuracy of the Ahlqvist and Damsten method was relatively high compared to some of the Kerley prediction formulae and profile method. This is similar to Kerley's own findings (Kerley 1965) in the original sample: the accuracy turned out to be better for younger specimens (under 30 years) than for older ones (30 years or more). In the opinion of this author, the application of histological methods is especially important for the determination of ages of 30 years or more, because for this age range there are only a few macroscopic characteristics useful for the determination of age.

In addition to Kerley's and Ahlqvist and Damsten's method, another histological ageing method developed by Singh and Gunberg in 1970, which utilizes the mandible, femur and tibia, has to be mentioned. As Ubelaker (1969) remarked: this method was based on a very restricted sample with most individuals between 50 and 75 years old. Moreover, Stout and Gehlert (1980) found that the error in age determination by this method ranged from 12 to 49 years. Singh and Gunberg chose to examine only two randomly chosen microscopic fields. Since considerable topographic variation in histomorphology occurs in cortical bone, considerable error may result from inadequate sampling of the histomorphology of the bone (Stout and Gehlert 1980).

With regard to the study of tibiae, a study by Ortner (1975) who investigated the ageing effects on osteon remodelling based on 101 human tibiae, has to be alluded to. He found a close relationship between the percentages of osteons and age.

In the present study the sample size was restricted to 20 individuals. Although the comparative studies of Stout and Gehlert (1980, 1982) were also based on only 20 and 13 individuals, in the opinion of this author a sample size of 20 individuals is on the small side. There are two reasons for using such a small sample size. Firstly, it was not easy to find 20 individuals with undamaged cortex and which had different ages between 20 and 90 years old. Secondly, in order to check the method and regression formula described and constructed by Ahlqvist and Damsten, the same number of individuals with about the same age range was employed.

It would be interesting to make a table of different regression formulae for the most important long bones, for example, the femur (see Ahlqvist and Damsten 1969), tibia (e.g. this vol.), fibula, humerus, radius and ulna. Regression formulae for the femur and tibia are summarized in Table 2. I plan to extend this table after having enlarged the sample size to 50 individuals. In the opinion of this author, such a table will help to make application of Ahlqvist and Damsten's method easier and more common.

One of the big advantages of microscopic methods over traditional gross morphological methods is the possibility of their application to badly fragmented bones. With the provisos that a cross-section can be obtained from the middle of the diaphysis and the specimen has an intact cortex, microscopic methods allow a determination of age (Kerley 1965).

Table 2 Regression formula for estimating age (y) from the percentage of osteons and osteon fragments (x)

Skeletal part	Regression formula	Standard error
Femur	$Y = 0.991X - 4.96$	± 6.71
(Ahlqvist & Damsten, 1969)		
Tibia	$Y = 0.989X + 3.31$	± 6.29
(this chapter)		

5 Conclusions

Microscopic methods are efficient and easily applicable procedures for the determination of skeletal age in human bones. The use of saw microtomes allows the preparation of thin (30 μm) and clear cross-sections (viz. anatomical structures in the sections remain intact). Decalcification, staining or embedding of the material is not necessary.

With respect to practical aspects, as well as to those of content, the method of Ahlqvist and Damsten (1969) is the method of choice. Application of Ahlqvist and Damsten's regression formula to 20 femora belonging to 20 individuals between 17 and 92 years old, led to good estimates of skeletal age. The standard error was 6.51 years. Application of the principle of Ahlqvist and Damsten's method, viz. counting the percentage of osteons and osteon fragments, to 20 tibiae also gave good results. The regression formula for the tibiae was of the following form: $y = 0.989x + 3.31 \pm 6.29$. Application of the same principle to other long bones as e.g. fibula, humerus, ulna and radius, would be very useful.

Acknowledgments. I would like to thank Dr G.J.R. Maat, Mr. W.J. Mulder, Mr. G.J. de Wolf and Dr F.W. Wilmink for their help during the study of the material and the preparation of this paper. I also wish to thank G.T. Hoogenberg for secretarial assistance, K. van Linschoten for preparing the drawings and P. van der Syde for making the photographs. The research was partly funded by a grant from the Leids Universiteits Fonds, Leiden.

References

Aeby C (1878) Das histologische Verhalten des fossilen Knochen und Zahngewebes. Arch Mikrosk Anat 15:371-382

Ahlqvist J, Damsten O (1969) A modification of Kerley's method for the microscopic determination of age in human bone. J Forens Sci 14:205-212

Bouvier M, Ubelaker DH (1977) A comparison of two methods for the microscopic determination of age at death. Am J Phys Anthropol 46:391-394

Frost HM (1958) Preparation of thin undecalcified bone sections by rapid manual method. Stain Technol 33:273-276

Kerley ER (1965) The microscopic determination of age in human bone. Am J Phys Anthropol 23:149-164

Kerley ER (1969) Age determination of bone fragments. J Forens Sci 14:59-67

Kerley, ER, Ubelaker DH (1978) Revisions in the microscopic method of estimating age at death in human cortical bone. Am J Phys Anthropol 49:545-546

Ohlah AJ (1977) Influence of microscopic resolution on the estimation of structural parameters in cancellous bone. In: Meunier PJ (ed) Bone histomorphometry. Fournie, Paris, pp 55-61

Ortner DJ (1975) Aging effects on osteon remodeling. Calcif Tissue Res 18:27-36

Schaffer J (1889) Uber den feineren Bau fossiler Knochen. Sitzungsber Akad Wiss Wien Math Naturwiss Kl 98:319-382

Singh JJ, Gunberg DL (1970) Estimation of age at death in human males from quantitative histology of bone fragments. Am J Phys Anthropol 33:373-382

Stout SD (1976) Histomorphometric determination of formation rates of archaeological bone. Calcif Tissue Res 21:163-169

Stout SD (1978) Histological structure and its preservation in ancient bone. Curr Anthropol 19:601-603

Stout SD, Gehlert SJ (1979) Histomorphological identification of individuals among mixed skeletons. Curr Anthropol 20:803-805

Stout SD, Gehlert SJ (1980) The relative accuracy and reliability of histological aging methods. Forens Sci Int 15:181-190

Stout SD, Gehlert SJ (1982) Effects of field size when using Kerley's histological method for determination of age at death. Am J Phys Anthropol 58:123-125

Stout SD, Teitelbaum SL (1976) Histological analysis of undecalcified thin sections of archaeological bone. Am J Phys Anthropol 44:263-270

Ubelaker DH (1969) Microscopic analysis of archaeological bone. In: Morse D (ed) Ancient disease in the Midwest. Reports of investigations 15. Illinois State Museum, Springfield, pp 66-72

Ubelaker DH (1974) Reconstruction of demographic profiles from ossuary skeletal samples: a case study from the tidewater Potomac. Smithsonian Contrib Anthropol 18

Ubelaker DH (1981) Approaches to demographic problems in the northeast. In: Snow DR (ed) Foundations of northeastern archaeology. Academic Press, New York, London, pp 175-194

Approaches to the Histological Age Determination of Cremated Human Remains

Susanne Hummel and Holger Schutkowski
Institut für Anthropologie
Georg-August-Universität
Bürgerstr. 50
W-3400 Göttingen
Germany

1 Introduction

From the late Bronze Age to the Migration Period cremation of the dead was the prevailing burial custom in middle and northern Europe. Thus, for a time span of about 50-60 generations, cremations are the major biological source for the reconstruction of historic population patterns and development. Proper age determination is one of the prerequisites for demographic reconstruction. The general state of preservation of a cremation (Fig. 1) shows clearly that an application of criteria for a classical morphological age determination, especially for adult individuals, is limited. The bones are highly fragmented, show cracks and are sometimes even deformed or twisted as a result of the combustion.

During the 1970s a consideration of these problems led to the suggestion of applying histological methods to cremated remains (Herrmann 1973, 1976). As far as the material is concerned this does not cause any problems since the necessary compact bone can be found in almost every cremation batch. Although the outer aspect of cremated bone shows considerable alterations due to the influence of temperatures above 700°C, the microstructural aspect is usually preserved. In the histological cross section one can distinguish different types of structural elements. It was shown that the type of structural element does not influence the amount of shrinkage, i.e. the shrinkage is linear (Fig. 2; cf. Hummel and Schutkowski 1986).

Thus, basic prerequisites could be warranted that consequently allow histological age determination in cremations. As a routine procedure this was done in our laboratory by means of qualitative estimations, that is, evaluating the age-dependent changes of the relative portions and appearances of the different histological microstructures. For that the experience acquired from native historical specimens was directly transferred to cremated remains. In order to minimize possible effects of different observer variations in evaluating a histological specimen we intend to check the applicability of histomorphometric methods to the burned material.

As far as histological age determination in native bone is concerned, different histomorphometric methods have been published (e.g. Ahlquist and Damsten 1969; Drusini 1987; Uytterschaut 1985). These methods are based on either the determination of osteons and osteon fragments per unit area or just counting the number of osteons. For that reason we attempted to develop a regression formula suited to cremated bones which followed these methods. These results will then be compared with qualitative histological age determinations.

2 Comparison of Histological Native and Cremated Specimens

Figure 3 shows the cross section of a native specimen of compact bone. The differentiation of non-Haversian canals, lamellar bone, osteons and osteon fragments is possible without any difficulty. In contrast, when the same specimen is experimentally cremated the distinction of the structural groups becomes more difficult (Fig. 4). The histological preservation of thin sections of historical cremations can vary markedly, ranging from aspects which are almost as clear as in native bone (Fig. 5) to aspects which are comparable to experimentally burnt specimens (Fig. 6). In such cases the use of polarized light does not lead to considerable improvement (Fig. 7) since, during the sinter process, the bone mineral becomes isotropic (Herrmann et al. 1990). The different aspects of historical specimens are certainly dependent on the temperature to which the body is exposed during cremation. If, in a historical cremation, a distinction of structural elements can easily be seen, most frequently these bones have been incompletely burnt and show traces of residual carbon.

3 Methods

The investigation for histomorphometric age determination on cremated bone was based on 18 dissection room specimens of known age at death. The ages of the individuals ranged between 19 and 76 years (mean age was 52 ± 17.6 years). First, thin sections of native bone of a thickness of approximately 80 μm were produced with a Leitz saw microtome. Then, we experimentally burnt a portion of the same bone in an oven at 1000 °C, for 1 h and, again, prepared thin sections.

Considering the variable appearances of historical cremated cross sections we decided to follow methods which avoided, as much as possible, the necessity of distinguishing the four structural groups as proposed by Kerley (1965). Thus, we used the Ahlqvist and Damsten method (1969) where only osteons and osteon fragments have to be distinguished. The area these structures covered in 1 mm^2 was evaluated and then expressed as a percentage. As a second approach, we followed a procedure proposed by Drusini (1987), where the number of osteons in 1 mm^2 was counted. We differed from those authors slightly in that we did not use an eye-piece with a grid, but reproduced the microscope picture on a TV-screen. From there we transferred the silhouettes of osteons and osteon fragments onto overhead transparencies, where a grid representing 1 mm^2 was positioned directly at the periosteal margin (Fig. 8). The sampling was carried out in anterior and posterior areas of the section for every specimen. Areas with extensive heat-induced cracks in the cross section were avoided. Using Ahlquist and Damsten's method those squares which were at least half-filled with osteons or osteon fragments were counted. In the case of Drusini's method this meant just counting the number of osteons. Only those osteons were counted which were at least visible for half of their total extension; half osteons at the edges of the grid were counted as half.

The results of these two approaches were compared with the data obtained from qualitative histological age estimations carried out on the same specimens. These estimations were based on the principles of age-related changes in compact bone as described by, for example, Amprino and Bairati (1938) and experience acquired with a reference series including specimens of known age.

Age allocation then followed a classification, which subdivided adult life into the categories 'adult' (20-40 years), 'mature' (40-60 years) and 'senile' (60+ years). The the first two categories were subdivided, according to Herrmann et al. (1990), into three further categories (young, middle and late). In order to get data comparable to the quantitative methods we transformed the allocation to an age class into years of life. If, for example, an individual was estimated 'late adult' this would mean an age of 36 years.

4 Results and Discussion

The results of the three different methods of histological age determination are listed in Table 1, and are briefly summarized here. According to Ahlquist and Damsten (1969) there exists a linear dependency between chronological age and the area covered with osteons and osteon fragments (Fig. 9). This is expressed by the regression formula R1: $y = 0.79x - 2.42$ with an SE of estimate of 14.5. The coefficient of correlation is 0.60.

The relation between chronological age and number of osteons (according to Drusini) also showed linear dependency (Fig. 10) as revealed by the regression formula R2*: $y = 1.41x + 3.88$ with a SE of estimate of 14.4. The correlation coefficient is 0.61.

Admittedly these results are not very encouraging. In the case of counting the number of osteons (Fig. 10), however, one has to mention that the specimen marked with an asterisk in Table 1 came from a 76 year old female individual who suffered from severe osteoporosis (see also fig. 11). In such cases with extensive resorption lacunae, counting the number of osteons must lead to a wrong conclusion, that is, the individual is aged too young. If the named case is omitted from the regression, the correlation becomes better (0.78). The regression formula would then be R2: $y = 1.79x - 11.41$ with an SE of estimate of 10.9.

Of all three methods applied, the best results were obtained with the qualitative histological age estimation (Fig. 12). The comparatively close correlation between chronological age and estimated age is reflected by a coefficient of 0.80. In this method it would not make sense to work with a regression formula since here it is not an abstract figure (like, for example, number of osteons) being correlated with the actual age but instead, age is determined directly by allocating the appearance of the cross section to an age class.

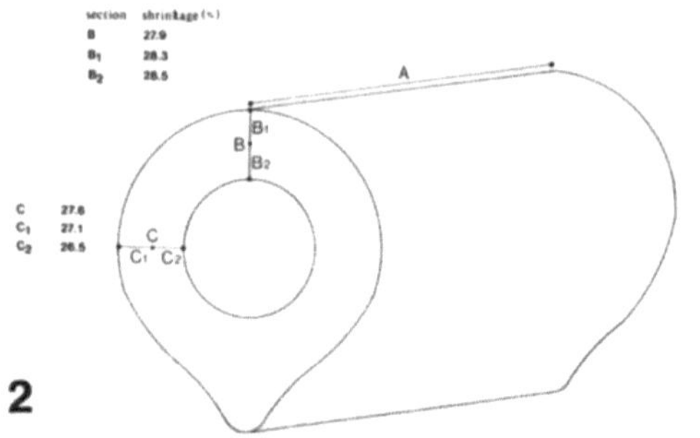

section shrinkage (%)
B 27.9
B1 28.3
B2 28.5
A
B1
B
B2
C 27.6
C1 27.1
C2 26.5
C
C1 C2
2

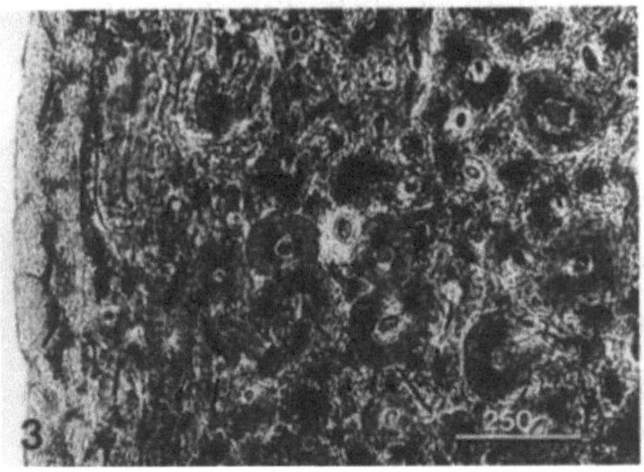

3
250

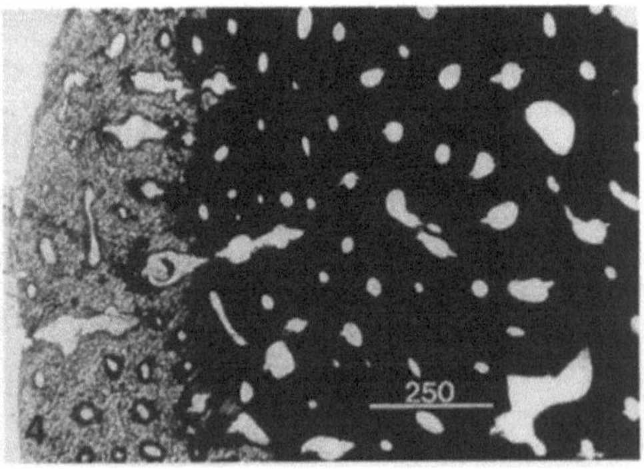

4
250

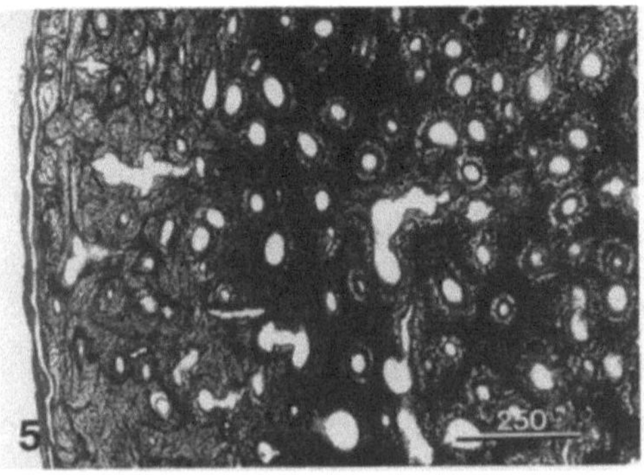

5
250

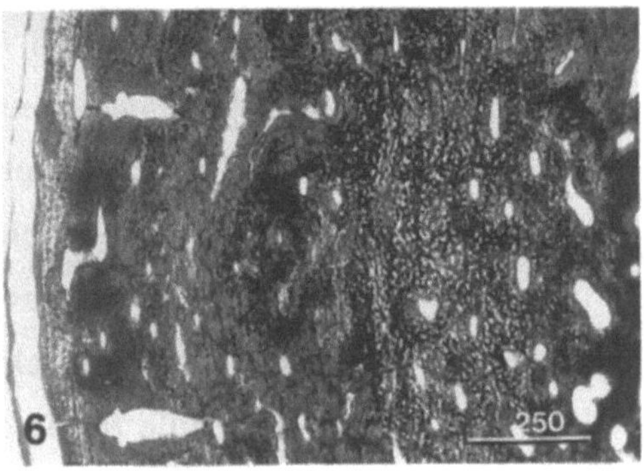

6
250

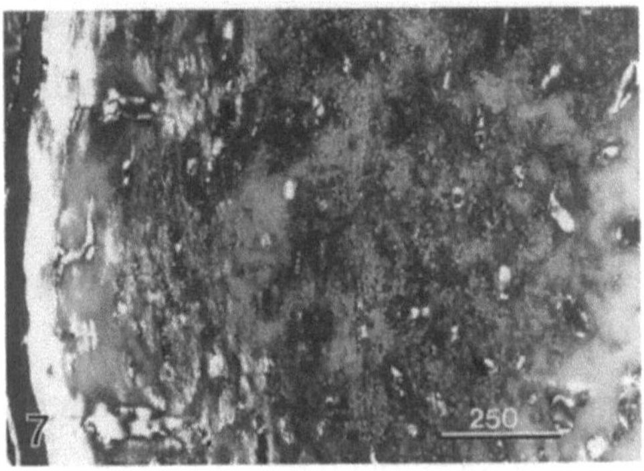

7
250

8

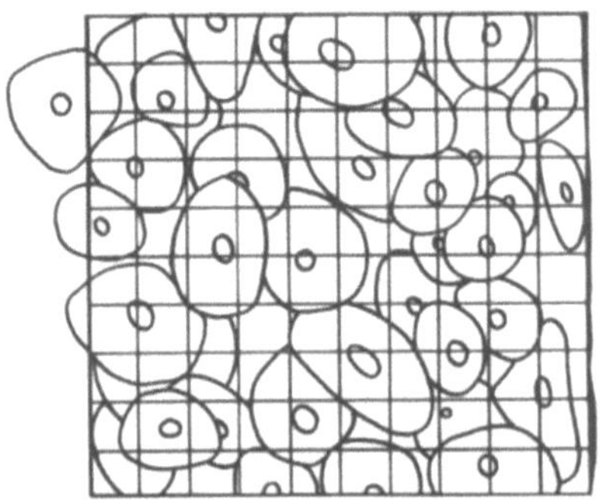

Fig. 1 General aspect of a well-represented historical cremation from the Bronze Age

Fig. 2 Shrinkage experiments revealed that there is no significant difference in shrinkage between the complete diameter of a cross section and and parts of it. This means that shrinkage along the cross section of compact bone is practically linear

Fig. 3 Thin section of native femur compacta (28-year old female). Non-Haversian canals, lamellar bone, osteons and osteon fragments are clearly distinguishable; scale in μm

Fig. 4 Thin section of experimentally cremated femur compacta. The specimen is taken from the same individual as in Fig. 3. Due to high temperature influence (1000°C) the bone microstructure alters. In the periosteal third distinction of structures becomes more difficult; scale in μm

Fig. 5 Cremated specimen from the Migration Period. The distinction of structures is almost as clear as in native bone. This is due to incomplete incineration; traces of residual carbon can be seen as dark areas in the middle of the compacta;
scale in μm.

Fig. 6 Cremated specimen from the Migration Period. This specimen is completely incinerated, which leads to an appearance very similar to experimentally burnt bone (cf. Fig. 4); scale in μm

Fig. 7 Same specimen as in Fig. 6. In completely burnt specimens polarized light does not usually improve the possibilities of structural distinction; scale in μm

Fig. 8 Example of a depiction of osteons and osteon fragments with a grid overlay that represents 1 mm^2. Such pictures served as the basis for the two histomorphometric methods (see text)

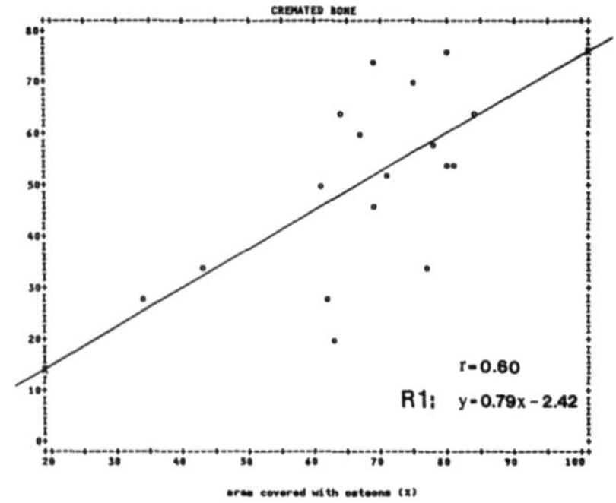

9

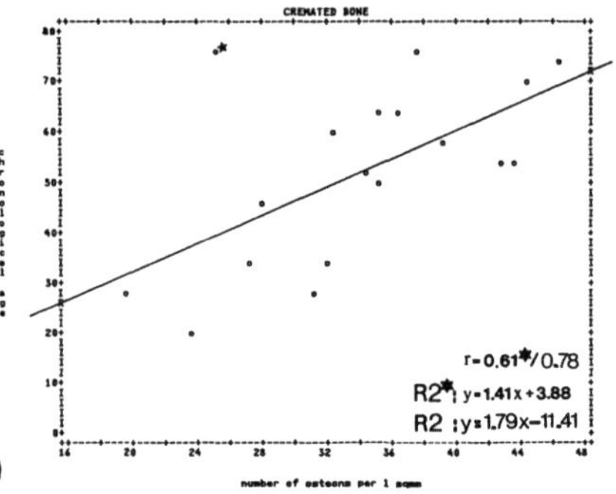

10

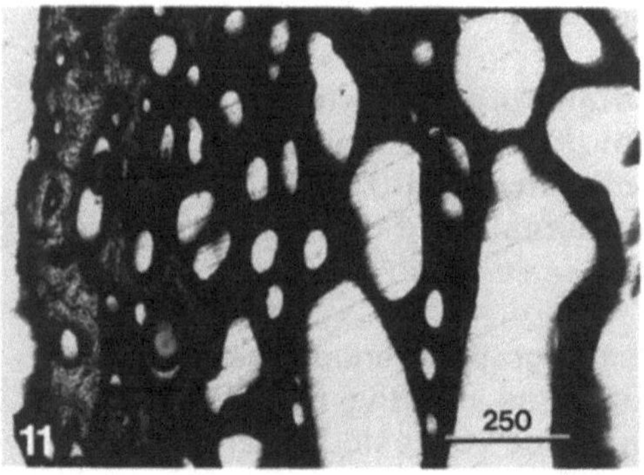

11

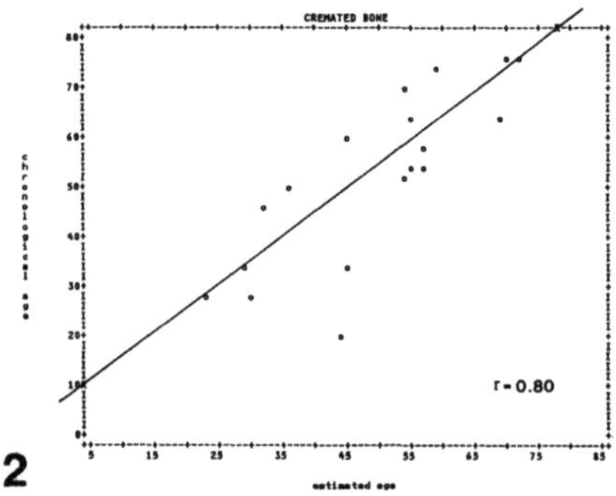

12

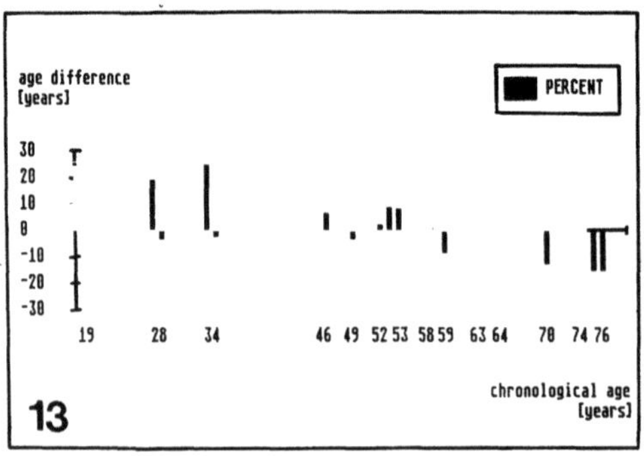

13

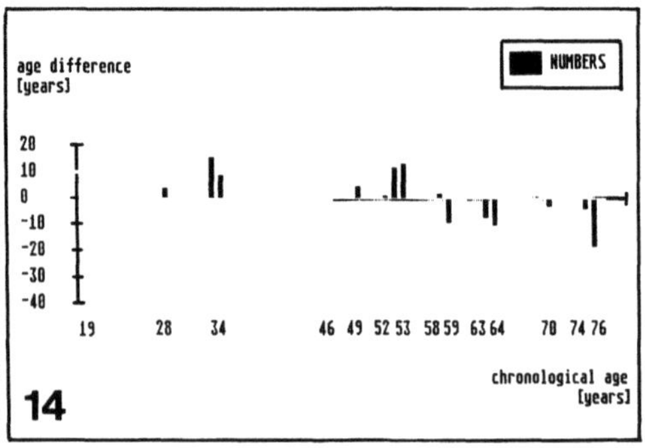

14

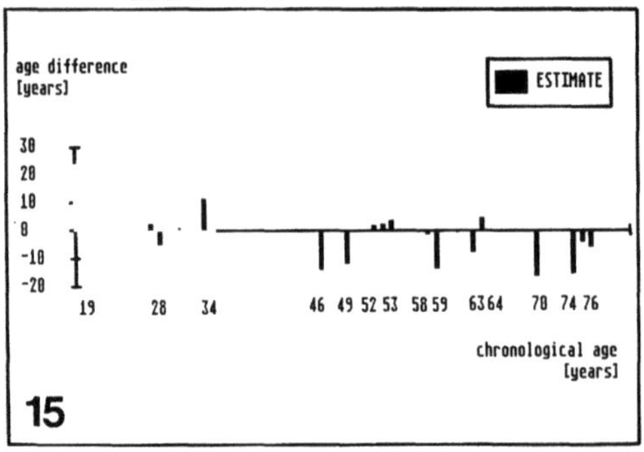

15

Fig. 9 Regression plot of "chronological age" against "area covered with osteons and osteon fragments"

Fig. 10 Regression plot of "chronological age" against "number of osteons per 1 mm^2". The regression formula y = 1.79x - 11.41 is based on all cases except the one marked with an asterisk. This equation is recommended if the age determination is to be carried out via the number of osteons (see text)

Fig. 11 Thin section of experimentally cremated femur compacta (76-year old female). This individual suffered from severe osteoporosis. Extended resorption lacunae clearly biased the results of histomorphometric methods. Using Ahlquist and Damsten's method an age of 60 years was obtained and using Drusini's method an age of 40 years. With a qualitative estimation, however, age was determined to be 72 years. Here the general appearance could easily be attributed to an elderly person

Fig. 12 Regression plot of 'chronological age' against 'estimated age'

Fig. 13 Deviation from chronological age when the regression given in Fig. 9 is used for age determination

Fig. 14 Deviation from chronological age when the upper regression given in Fig. 10 is used for age ddetermination

Fig. 15 Deviation from chronological age when the age determination is based on a qualitative estimation

Table 1 Comparison of determined ages with chronological age

Chronological age	Calculated age using R1 (cf. Fig. 9)	Calculated age using R2 (cf. Fig. 10)	Estimated age (cf. Fig. 12)
19	47	37	44
28	47	48	30
28	24	31	23
34	58	49	45
34	32	43	29
46	52	44	32
49	46	53	36
52	54	52	54
53	61	64	55
53	61	66	57
58	59	59	57
59	50	49	45
63	48	55	55
64	64	53	69
70	57	66	54
74	52	69	59
76	61	57	70
76*	60	40	72

R1 refers to the regression obtained when the age determination is based on the percentage of area of osteons and osteon fragments; R2 to counting the number of osteons; and "estimated age" to a qualitative evaluation.

* Indicates specimen from 76-year old female who suffered from severe osteoporosis

Nevertheless, the regression formula which can be calculated for 'chronological age' against 'estimated age' is given here in order to make the three methods comparable from a formal point of view. The expression is $y = 0.97x + 4.76$ with a SE of estimate of 10.8.

We then examined in which direction the calculated or estimated ages deviated from chronological age. Figures 13-15 show that all three methods are systematically biased in so far as young individuals tend to be aged too old and vice versa. This effect appears to be least pronounced with the age estimations. A slight bias, however, might be expected anyway, because at the extreme ends of the age range wrong diagnosis of an individual forces it in one direction: a young individual cannot be judged younger than "young", and an old individual similarly can respectively not be judged older than "old".

Corresponding to the results given above the calculation of the mean differences between chronological age and determined age led to the best result in the qualitative method. The mean difference is 8.4 years $\pm$ 6.5. By the quantitative methods of Ahlquist and Damsten it is 10.9 years $\pm$ 8.6 and by Drusini's method it is 10.5 years $\pm$ 8.9.

The comparison of quantitative, histomorphometric methods with a qualitative method, which is based on evaluating the relative portions, size and shape of the different types of bone microstructures revealed the qualitative method to be preferable. At first sight this seems to be an unexpected and astonishing result, because the general prerequisites, such as linear shrinkage of the cross section (cf. Fig. 2) and the feasibility of distinguishing microstructures are basically given in experimentally cremated specimens (Fig. 4) as well as in historical cremated specimens (Figs. 5 and 6). However, difficulties do occur in cremated specimens when the applicability of a method is dependent on a very precise distinction of single structural elements, as is necessary in both quantitative methods, especially Ahlquist and

Damsten's method. Of course, the qualitative assessment of age generally has to deal with these difficulties too. This is also reflected by the fact that from our experience a qualitative age determination reveals better results in native bone than in cremated bone. That for cremated specimens, however, a qualitative approach yields the best results of the three methods considered may be due to the fact that the general aspect of position, size and shape of microstructures is more important for evaluation than a precise distinction of structures. In addition, with a qualitative estimation it is advantageous that the aspect of the whole cross section is evaluated, whereas quantitative methods are restricted to a certain visual field in the periosteal third.

Within both quantitative methods the one which is less dependent on a precise distinction of structures, i.e. counting osteons according to Drusini, is clearly preferable. This, however, only holds if the obvious pathological exception (as discussed earlier) is omitted from the regression.

Summarizing the findings we recommend that it is best to diagnose the age of cremated historical specimens by means of a qualitative evaluation of age dependent changes in bone microstructure. We concede that this recommendation can bear difficulties in so far as it requires considerable experience and continuous study of the subject. Those who are less familiar with the morphological appearance of thin sections from cremated bone and therefore prefer to use a method less dependent on experience may use Drusini's method. In this case we recommend use of the regression formula which was calculated after omitting the obvious pathological specimen ($y = 1.79x - 11.41$).

References

Ahlquist J, Damsten O (1969) A modification of Kerley's method for the microscopic determination of age in human bone. J Forens Sci 14: 205-212

Amprino R, Bairati A (1938) Le variazioni nella struttura dell'ossa in relazione all'et e la loro importanza medico-forense. Arch Anthropol crim, Psichiatr Med Leg 16:61-74

Drusini A (1987) Refinement of two methods for the histomorphometric determination of age in human bone. Z Morphol Anthropol 77:167-176

Herrmann B (1973) Möglichkeiten histologischer Untersuchungen an Leichenbränden. Mitt Berliner Ges Anthropol, Ethnol Urgesch 2:164-167

Herrmann B (1976) Experimentelle und theoretische Beiträge zur Leichenbranduntersuchung. Homo 27:114-118

Herrmann B, Grupe G, Hummel S, Piepenbrink H, Schutkowski H (1990) Prähistorische Anthropologie. Leitfaden der Feld und Labormethoden. Springer, Berlin Heidelberg New York

Hummel S, Schutkowski H (1986) Das Verhalten von Knochengewebe unter dem Einfluß höherer Temperaturen. Bedeutung für die Leichenbranddiagnose. Z Morphol Anthropol 71:1-9

Kerley ER (1965) The microscopic determination of age in human bone. Am J Phys Anthropol 23:149-163

Uytterschaut HT (1985) Determination of skeletal age by histological methods. Z Morphol Anthropol 75:331-340

Comparative Histological and Microradiographic Investigations of Human Bone

Friedrich W. Heuck
Hermann-Kurz-Str. 5
W-7000 Stuttgart
Germany

Bone is a complex hard tissue and a highly organized connective tissue constituting the main calcified component of the skeleton in vertebrates. The mechanical properties of bone, its hardness, rigidity and strength are a result of the composition of its matrices. These are a dense network of organized collagen fibres and a mineral phase, which is primarily composed of poorly crystallized calcium phosphate, later changing into hydroxyapatite and some non-collagenous proteins. The nature of the compact bone tissue of the cortex of limb bones has been studied in many types of vertebrates. This part of the bone has to carry and take up the main forces acting upon the skeleton, in particular the weight of the body.

Our knowledge about the metabolic processes of the hard tissue bone is incomplete and fragmentary. Bone transformation under normal and pathological conditions is not fully understood. On the macromorphological level, X-ray studies of bone have contributed immensely to our scientific knowledge about bone disorders. Since bone is a highly mineralized connective tissue, on the microscopic level, microradiographic studies of mineral distribution in comparison with the stained bone sections are most informative.

1 Remarks on Methodology

It is taken for granted that the general principles of bone microradiography are well known, thus only some remarks on methodology will be presented. Chemically, bone can be divided into inorganic and organic fractions. The organic fraction, including water, represents at least half of the fresh weight of bone. The volume occupied by this fraction is about 70% and the content of water varies with age of the bone and the state of growth. In newly forming bone the water content may be as high as 70% and in old, compact bone as low as 10%. The inorganic substance of the bone consists mainly of hydroxyapatite ($Ca_{10}(PO_4)_6(OH)_2$), calcium phosphates or calcium carbonate and, in smaller amounts, other ions.

The fresh bone specimens can be dehydrated and then stained in a fuchsin alcohol solution for about 4 days (Schenk 1965; Heuck 1969; Schenk et al. 1969). Pieces of trabecular bone or smaller pieces of compact bone can be embedded into methylmethacrylate, which penetrates into the specimen, before cutting and grinding. Several types of sawing machines have been developed (Heuck 1969; Uytterschaut 1985). Our laboratory uses a Leitz saw microtome which allows direct cutting of 10-100 μm thick sections with plane parallel surfaces. However, in order to secure good microradiographs, the bone sections must be ground and polished. This is done with fine grained waterproof sandpapers and uses water as a wetting agent. Final polishing is done between two plain frosted glass plates with extremely fine surfaces.

In order to select X-ray energies suitable for the imaging of microscopic sections of bone tissue it is necessary to bear in mind that the K absorption edge of calcium is 3.06 Å. Provided that suitable X-rays for microradiography are chosen, the major part of the absorption will indicate the distribution of mineral salts (Engström

1970; Jowsey 1977a,b). Independent of the thickness of bone sections, poorly mineralized areas, such as osteoid seams, can be identified only in microradiographs produced with X-rays of appropriate wavelength.

Therefore, our laboratory mainly uses a Phillips CMR 5 (Neumann 1958; Heuck 1969). The X-ray tube has a beryllium window, 50 μm in thickness, and a 0.3 mm x 0.3 mm focal spot. The voltage used was 4.5-5.0 kV, so the radiation had a wavelength limit of 2.7 Å and the beryllium window in the tube absorbing the radiation a limit of 4Å. Within the range of this spectrum falls the K absorption edge of the calcium at 3.06 Å; therefore, mineral deposits are easily detectable.

For quantitative measurements a reference system, for example, an aluminium step-wedge, should be microradiographed simultaneously with the specimen. The microradiograms are prepared on fine grained photographic emulsions, such as Kodak high resolution spectroscopic plates 649 or Lipman emulsion (Neumann 1958; Jowsey 1977a,b). The exposed plates are developed in high-resolution developer and processed in the usual way. The film and stained ground sections of bone are mounted with Eukitt in corresponding positions on the same microradiographic slide (Heuck 1969, 1971, 1974, 1976). Now, comparison of the corresponding regions of the bone will be possible.

2 Normal Bone

During the growth of an individual the pattern of bone structure changes. Thus, one can easily differentiate the bone tissue of a neonate, a child or an adult, if investigating the same area of the skeleton. During the phase of intense bone transformation, which happens in childhood, many poorly mineralized osteons and many resorption lacunae can be observed side by side. In the normal bone of adults they are very rare. The mineral concentration in the osteons is mainly equal and small differences in the mineralization of the osteons and the intermediate lamellae of normal adult bone tissue can be observed. Microradiograms of normal compact bone demonstrate that the Haversian systems have a varying degree of X-ray absorption, indicating different degrees of mineralization. The young Haversian systems have the lowest concentrations of bone mineral. Some examples of bone structure and mineralization in the cortical area of compact bone of the proximal diaphysis of the femur are illustrated in Fig. 1. The pattern of embryonic or newborn human bone, the growing bone from a 4-year-old boy and human adult bone are different. In some fields the Haversian systems are limited by higher mineralized cement lines than the osteon itself. The lacunae of the osteocytes are well demarcated, the canaliculi are visible in the stained sections only (Heuck 1974, 1976).

In trabecular bone, areas of low and high mineralization vary in a more irregular way than in compact bone. The less mineralized areas in normal trabecular bone are the younger ones. Cancellous bone tissue is composed of a wide meshed network of thin, irregular rods or plates, which are more or less equally mineralized (Fig. 2). Between the trabeculae relatively large spaces are found which are filled with bone marrow. There are few osteons in the trabeculae, and those that do occur are atypical. Sporadically, resorption lacunae are observed in normal trabecular bone tissue. The rare cement lines are frequently highly mineralized and are well demarcated in adult bone. Sometimes a higher mineralization in the interstitial lamellae are very rarely found around a single bone cell or osteocyte in normal bone tissue. These areas of the hard tissue deserve special attention.

3 Pathological Bone

In our group we have prepared sections to investigate and analyse more than 600 bone specimens representing different diseases in humans and also in animals. Therefore, I shall illustrate some examples of atypical or pathological patterns of bone mineralization in different groups of diseases; these are metabolic bone diseases, genetic disorders, bone tumours and bone metastases.

We have observed remarkable differences in mineral concentration in well-defined areas of bone tissue (Heuck 1976; Jowsey 1977a). Sometimes the stained areas coincide with low mineralization, for example the osteoid seams in rickets (Fig. 3). In other diseases the highly mineralized zone, for example, a cement line, barely shows on staining. This is the situation in marble bone disease where by using diffraction methods, we found very closely packed and extremely small amorphous particles of mineral which are responsible for the high density.

In hormonal or metabolic bone diseases the exchange of mineral and/or the transformation rate are abnormal (Jowsey 1977a). The patters in primary or secondary hyperparathyroidism are quite different from that of normal bone; the mineral distribution is very irregular (Fig. 4). Highly mineralized areas are seen side by side with poorly mineralized bone tissue. Remarkable are the many places of periosteocytic demineralization, which occasionally join the large areas of low mineralization. These findings are consistent with the hypothesis of osteocytic osteolysis as the first steps of bone resorption.

In Paget's disease the histological patterns of the so-called mosaic pattern are typical. In the corresponding microradiographs one can see the irregular bone mineral concentration in these areas of transformation (Fig. 5). Sometimes the finding of periosteocytic demineralization of Paget's disease is very similar to that in secondary hyperparathyroidism. The increase of highly mineralized cement lines is remarkable in this particular case.

A typical example of a genetic disorder is marble bone disease. Histological and microradiographic studies demonstrate great differences in the pattern of mineral concentrations. In some areas large differences in mineral concentration are apparent (Heuck 1971, 1976). In a special example, the periosteocytic low mineralization can be observed adjacent to a comparatively high mineralization in the cement lines, and this sometimes occurs around a single osteocyte! These findings are the result of regional differences in the mineral metabolism of the osteocytes in marble bone disease.

Bone-forming tumours, for example, osteoid osteomas and osteosarcoma, can develop different mineral patterns of new bone. In an osteoid osteoma many small osteons with broad osteoid seams in the newly formed bone can be seen, as can irregular bone mineralization and areas of different lacunae of osteocytes, both normal and enlarged (Fig. 6). The new bone in osteosarcoma is woven bone with mineralization defects, many Howship's lacunae and different sizes of osteocytes.

Bone tissue in parosteal osteosarcoma shows a much regular pattern and a uniform distribution on mineral in the newly formed tumour bone. This is called woven bone and it contains some new osteons between the lamellar bone tissue.

The reaction of bone tissue to several types of cancerous tissue in instances of metastases shows a quite different histological picture (Heuck 1976). Bone reactions in carcinomas of the breast are totally different. In some cases the remineralization and/or reossification of destroyed bone shows new bone formation of different qualities: woven bone, lamellar bone and sometimes totally irregularly formed new bone tissue appear mixed up with tumour tissue. In these cases the mineral pattern in histological and microradiographic images demonstrates the "battle" between tumour tissue and bone tissue. Howship's lacunae, periosteocytic mineral defects, and osteoid seams can be found side-by-side.

Osteoblastic metastases are found in carcinoma of the stomach. Histological and microradiographic investigations have demonstrated the findings of the so-called bone-in-bone (Fig. 6b). This means in the marrow spaces the cancerous tissue has introduced new woven bone formation. The mineralization of this new bone is irregular. These findings are typical in gastric carcinoma. In the case of osteoblastic metastases in prostatic carcinoma, the new bone formation, again, is very different and much more irregular. Highly mineralized bone, broad osteoid seams and the so-called buried osteoids are found in the same region.

4 Quantification

Electronic methods for analysis have been developed. The square dimension of bone tissue and the mineral concentration per unit volume in different bone areas can be obtained, and the number of bone cells or osteocytes can be counted. Computerized methods may specify the mineral concentrations on a selected measuring line or in a small area of bone. The determination of histograms is possible.

5 Conclusion

The only tissue which is preserved beyond the lifetime of a human or animal is bone! Therefore, this mineralized tissue may preserve many facts about the life history of an individual, human or animal.

The ionic exchange in bone after death may be different in highly mineralized versus poorly mineralized bone tissue (Garland 1989; Piepenbrink 1989). Attention should be paid to the different solubilities of amorphous and/or crystalline deposits of bone mineral both in high density and low density areas of bone tissue.

Knowledge of the typical mosaic patterns of mineral concentration found in normal and pathological cortical and trabecular bone during the decades of life should make the interpretation of histological sections of prehistoric bone easier.

Fig. 1 Microradiographs of normal bone tissue (proximal diaphysis of the femur).
a Newborn boy, 7 months and 18 days old.
b Boy, 4 years and 2 months old. Note the differences in mineral concentration and areas of bone transformation.
c 49-year old male. Note the pattern of distribution and concentration of mineral in the Haversian system

Fig. 2 Examples of microradiographs of trabecular bone (lumbar vertebrae) of a 58-year-old male. Small differences in the mineral concentration in the lamellae of the trabeculae can be seen

Fig. 3 Generalized bone disease.
a Osteomalacia in a 56-year-old female. Low mineralization of the osteoid seams (arrows) and higher mineral concentration in the so-called mineralization front.
b Marble bone disease in a 43-year-old female. Highly mineralized areas: periosteocytic and cement lines

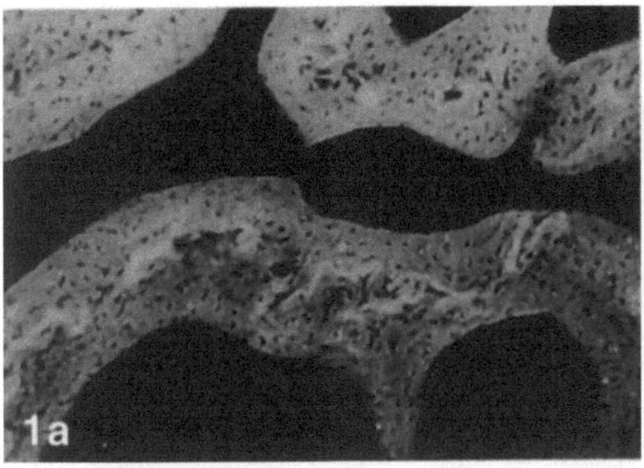
1a

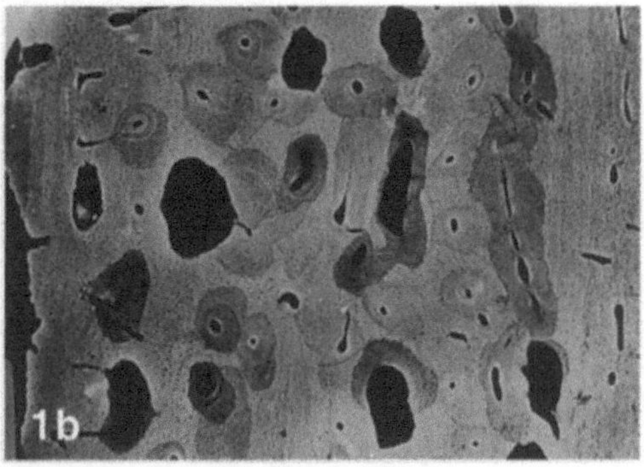
1b

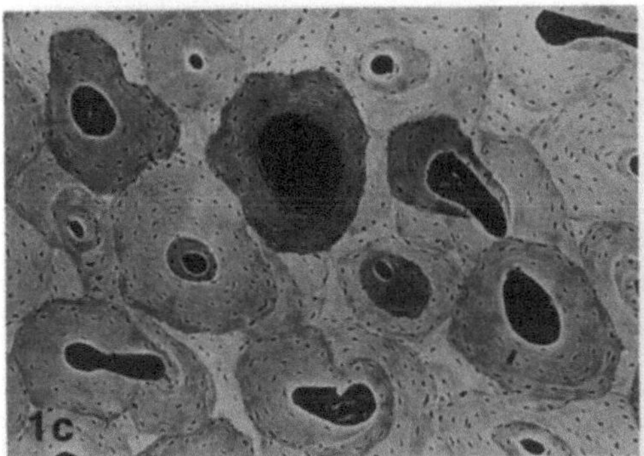
1c

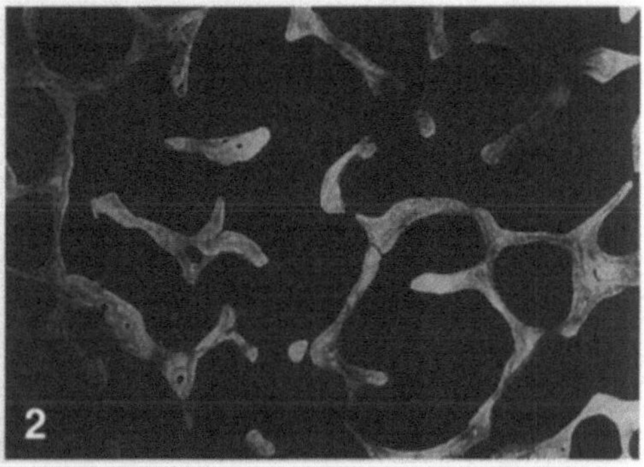
2

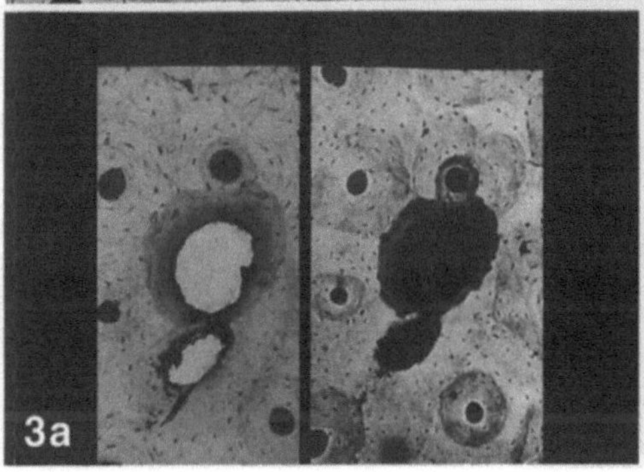
3a

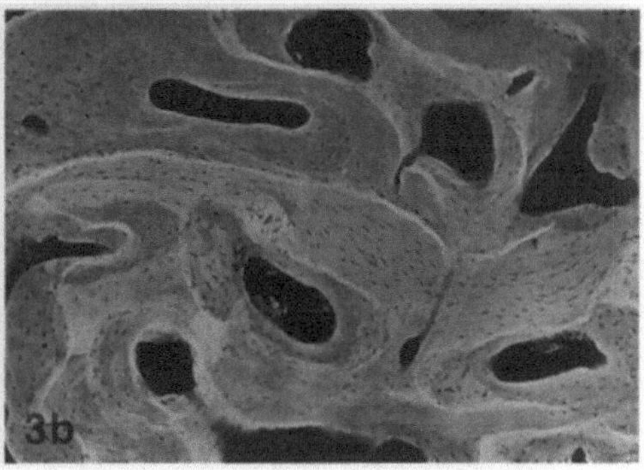
3b

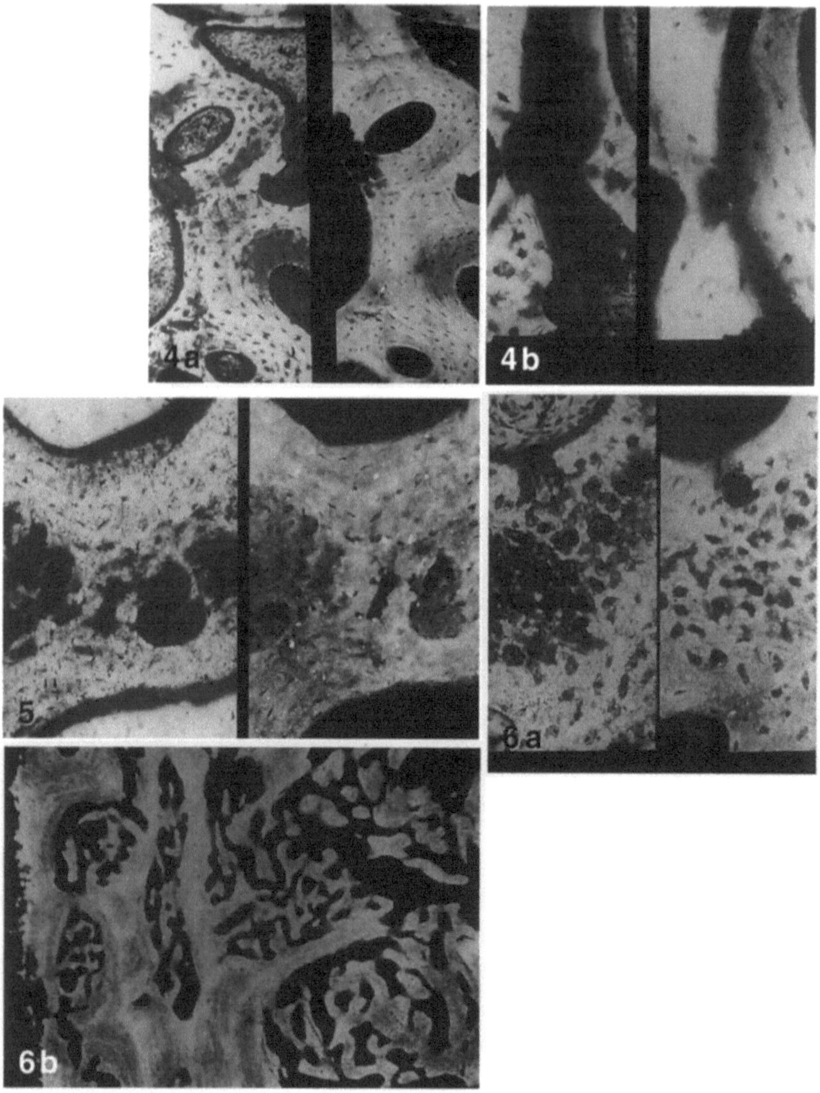
4a
4b
5
6a
6b

Fig. 4 Examples of very irregular patterns of bone mineralization in renal osteodystrophy (secondary hyperparathyroidism) in an 18-year-old male.

a Poor mineralization, periosteocytic osteolysis and Howship's lacunae in the same area; different magnifications.

b Same patient: "buried osteoid seams" and periosteocytic demineralization

Fig. 5 "Mosaic pattern" of compact bone in Paget's disease; 57-year-old male. Periosteocytic demineralization and osteolysis; Howship's lacunae, differences in mineral concentration.

Fig. 6 Examples of bone tumours.

a Osteoid osteoma in the proximal diaphysis of the tibia of a 17-year-old boy. The bone sample was taken from the newly formed bone. The broad osteoid seams in the young Haversian systems and the big lacunae of the osteocytes are remarkable.

b New bone formation (bone-in-bone) in the area of metastasis in a 50-year-old male with carcinoma of the stomach. Low mineral concentration in the newly formed bone tissue.

References

Engström A (1970) Microradiography of normal bone. In: Diëthelm L, Obssou O, Struad F (eds) Handbuch der medizininschen Radiologie, vol. 1V/1. Springer, Berlin Heidelberg New York, pp 296-316

Garland AN (1989) Microscopic analysis of fossil bone. Appl Geochem 4:215-229

Heuck F (1969) Mikroradiographische Untersuchungen der Mineralisation des gesunden und kranken Knochengewebes. Radiologie 9:142-154

Heuck F (1971) Investigations of high density areas in metabolic bone disease. Isr J Med Sci 7:477-480

Heuck F (1974) Mikroradiographie. Vehr Dtsch Ges Pathol 58:114-134

Heuck F (1976) Allgemeine Radiologie und Morphologie der Knochenkrankheiten. In: Diëthelm L, Heuck F, Obssou O, Ranniger K (eds) Handbuch der medizinischen Radiologie, vol. V/1. Springer, Berlin, Heidelberg New York, pp 3-303

Jowsey J (1977a) Metabolic diseases of bone. Saunders, Philadelphia

Jowsey J (1977b) The bone biopsy. In: Avioli, LV (ed) Topics in bone and mineral disorders. Plenum, New York

Neumann KH (1958) Anwendung von Röntgenlicht für histochemische Untersuchungen. In: Granmann W, Neumann KH (eds) Handbuch der Histochemie. Fischer, Stuttgart, pp 339-399

Piepenbrink H (1989) Examples of chemical changes during fossilization. Appl Geochem 4:273-280

Schenk RK (1965) Zur histologischen Verarbeitung von unentkalkten Knochen. Acta Anat 60:3-19

Schenk RK, Mertz WA, Müller J (1969) A quantitative histological study of bone resorption in human cancellous bone. Acta Anat 74:44-53

Uytterschaut HT (1985) Determination of skeletal age changes by histological methods. Z Morphol Anthropol 75:331-340

Histomorphometric Methods Applied to Bone

Georges Boivin and Pierre J. Meunier
Institut National de la Santé et
de la Recherche Médicale, U 234
Pathologie des Tissues Calcifies
Faculté Alexis Carrel
Rue Guillaume Paradin
F-69372 Lyon Cedex 08
France

1 Introduction

Histomorphometry, or quantitative histology, consists of counting or measuring tissue components: the cells or the extracellular constituents or both. Because of the well-architectured three-dimensional organization of bone tissue, because this tissue is continuously remodelled throughout life, and because most types of bone diseases are anatomically characterized by a quantitative abnormality, this quantitative approach is particularly advantageous for the analysis of age-related physiological bone changes and of the pathophysiology of bone diseases. Indeed, bone histomorphometry applied to well-preserved undecalcified samples permits the measurement not only of static parameters, such as the bone volume, the imprints of previous bone remodelling events and the number of cells, but also of dynamic parameters, by the use of the tetracycline double labelling procedure (Frost 1969). This process has permitted the introduction of the dimension of time into the quantitative analysis, thus providing access to vital information on organ, tissue and cell turnover kinetics.

In the last three decades, the field of bone histomorphometry has progressed markedly, and now serves two main needs; namely, the diagnosis of bone disease and research devoted to achieving a better understanding of the pathophysiology of bone disease and of the histological effects of new therapeutic approaches (Eriksen 1986). In human and animal studies, bone histomorphometry is the only method suited to evaluate tissue and cell changes at the level of the intermediary organization of bone (Frost 1983), that is the osteon in cortical bone and the basic structural unit (BSU) or cancellous packet in spongy bone (Frost 1973; Meunier 1983). Measurements of biochemical markers of bone remodelling and non-invasive methods of bone mass measurements are unable to give access to this BSU level. Hence, a local system of specialized bone cells and their precursors (basic multicellular unit, BMU) is responsible for the remodelling process which culminates in a BSU.

Bone histomorphometry demands strict methodological conditions because an unequivocal qualitative identification and description of the parameter to be measured is pivotal to any quantitative analysis. Thus, bone biopsy techniques yielding a well-preserved tissue sample and histological methods allowing discrete variables to be measured are two prerequisites for accurate bone histomorphometry.

Concerning the main methods used for the preparation of bone samples, the techniques generally employed to study bone tissue, and the histomorphometric approach to bone histology, an extensive bibliography will be found in Meunier (1973, 1977, 1983), Jowsey (1977), Jaworski (1976, 1982), Jee and Parfitt (1981) and Dickson (1984).

2 Bone Biopsy

Iliac biopsy under local anaesthesia is now universally used because it does not require any surgical protocol, provides both cortices with a large area of spongy bone, and is safe. Indeed, a multicentre survey on the side effects of 14 810 iliac biopsies has been reported (Duncan et al. 1981), 9030 being obtained by the transiliac horizontal approach through the iliac crest. An overall incidence of complications of 0.52% was found. The most common problems were haematomas (0.24%), pain for 7 or more days (0.11%), femoral neuropathy (0.09%), and skin infection (0.07%). Eleventh rib biopsy, initially used by Epker and Frost (1965), has now been abandoned in human adults because it necessitates orthopaedic surgery and provides little spongy bone, with a risk of pleural damage.

In the iliac bone, the bone biopsy may be obtained either by a vertical approach, as described first by Sacker and Nordin (1954) or by a horizontal transiliac method, as described by Bordier et al. (1964). The latter technique gives a specimen with cortex at both ends which is a representative volume of iliac bone in its entirety and permits the measurement of the total bone density (Fig. 1). Moreover, a large area of spongy bone and the compact bone of two cortices can be analysed separately. The transiliac biopsy is taken from an area situated 2 cm below the summit of the iliac crest and 2 cm behind the antero-superior iliac spine. It requires local anaesthesia not only of the external but also of the internal periosteum, and a 1 cm long skin and muscle incision to be made. In order to reduce sampling variations we use a manual trephine having a 7.5 mm inner diameter and, when possible, we take and measure on two contiguous cores which are embedded in the same block of tissue.

The experience of the operator is an important determinant of the quality of the specimen. About 16.7% of biopsies were poor when the operator was inexperienced, whereas 8.8% were unsatisfactory in experienced hands (Duncan et al. 1981). In our laboratory, of 390 transiliac biopsies which were performed by an experienced operator 64% were complete and well preserved, and only 7.4% did not allow a valid diagnosis (Chavassieux, unpublished data).

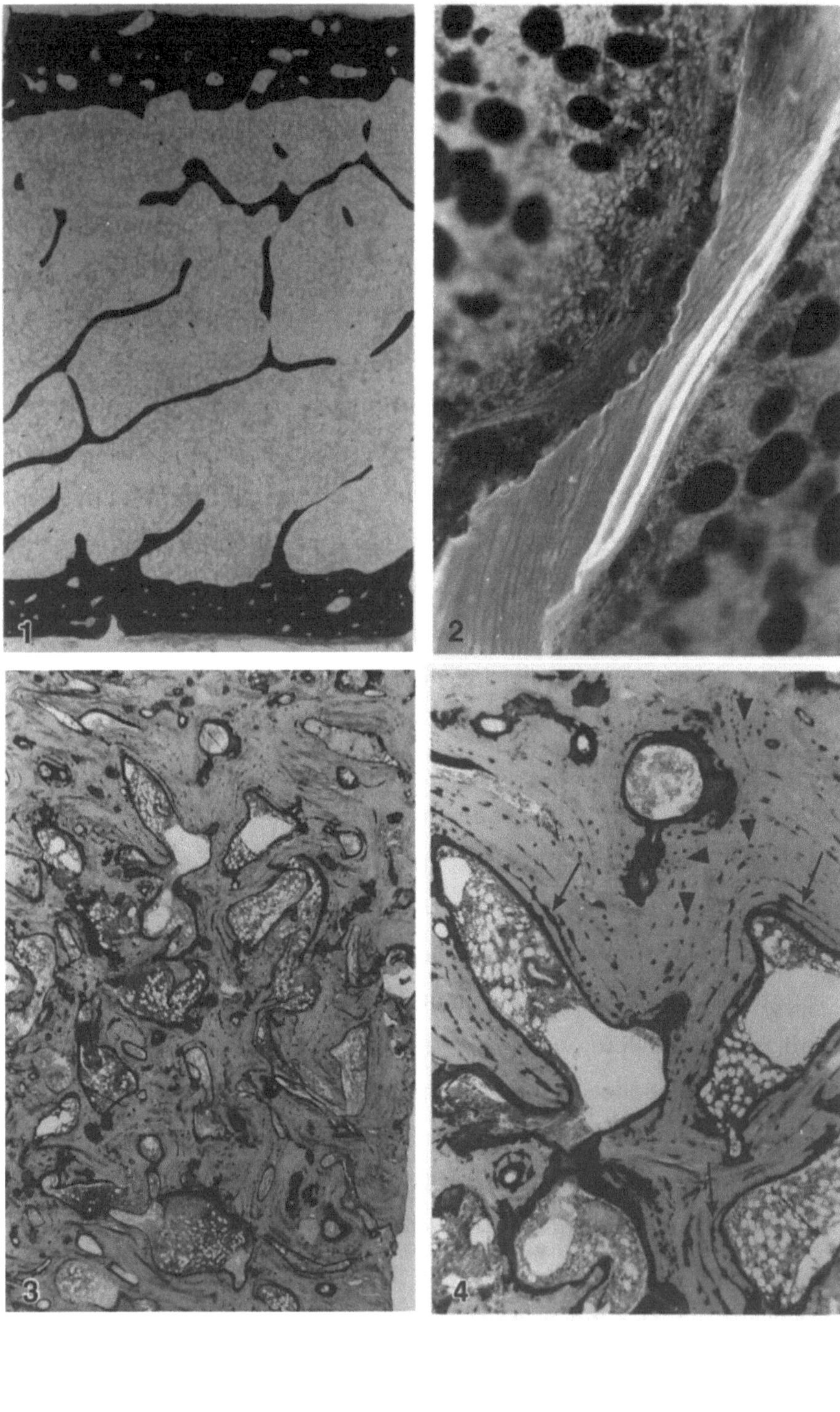
1
2
3
4

Fig. 1 Section (solochrome cyanin R staining) of an undecalcified transiliac bone biopsy in a case of osteoporosis. Between the two cortices (compact bone), cancellous bone of trabeculae is rarefied

Fig. 2 Unstained section of an undecalcified transiliac bone biopsy in a case of secondary hyperparathyroidism. Tetracycline double- and single-labels are observed (with ultraviolet light) along trabecular surfaces

Fig. 3 Illustrates the histological osteosclerosis due to an increase of cancellous bone volume in a section (Goldner's trichrome staining) of an undecalcified transiliac bone biopsy in a case of skeletal fluorosis

Fig. 4 Shows clearly the presence of linear formation defects (arrows) and mottled bone with mottled periosteocytic lacunae (arrowheads); a section (Goldner's trichrome staining) of an undecalcified transiliac bone biopsy in a case of skeletal fluorosis.

3 Histological Techniques

It is mandatory that bone samples be processed without prior decalcification. The most commonly used fixative is 80% ethanol which does not dissolve tetracycline. Embedding bone samples in plastic monomers provides blocks of the same hardness as the bone, and methylmethacrylate is the most convenient plastic (Jowsey et al. 1965; Meunier 1983; Schenk et al. 1984). Non-consecutive serial sections, 4 - 20 μm thick, are cut with specially designed microtomes (formerly Jung K type and to date Reichert Polycut type) equipped with tungsten carbide-edged knives.

Multiple staining procedures are available: the only prerequisite is to use a method which allows the unequivocal identification of osteoid and of bone cells. Solochrome cyanin R (1% aqueous solution) is a reliable osteoid stain, having the same discriminant capacity as Von Kossa's procedure, recognized as the reference method (Meunier et al. 1975). Solochrome cyanin R is also quite convenient for measurements of bone volume with the use of image-analysing computers because it produces a clear contrast between dark-stained bone trabeculae and almost unstained marrow spaces. In contrast, the widely used Goldner's trichrome underestimates both osteoid volume and osteoid surfaces, though it is excellent for evaluating resorption parameters.

The analysis of tetracycline double-labelling is possible as a routine procedure on unstained bone sections that are 15 - 20 μm thick (Fig. 2). A convenient labelling protocol to use before biopsy consists of giving a patient 10 - 15 mg/kg/day dimethylchlortetracycline (DMC) orally for 2 days, no medication for 12 days, and DMC again for 4 days. The bone biopsy must be taken 4 - 7 days after the end of the second labelling period.

4 Measurements and Histomorphometric Parameters

Different methods can be used for the histological measurements performed on sections of bone samples. Both are based on Delesse's principle which permits deduction of three-dimensional numerical information from two-dimensional images (Delesse 1847).

The first method consists of projecting integrating grids over the section of bone sample. The percentage of hits overlying a structure or the number of intersections between the grid and the bone perimeters are accurate estimates of the volumetric or surface density of the analysed component. This manual point-counting system (Schenk et al. 1969; Meunier 1983) is still the most commonly used although it requires analysis of a large number of fields and is tedious and time-consuming.

The second method uses an image-analysing computer (Meunier 1973) which reduces analysis time considerably. These fully automated devices are well suited for the measurement of bone volume or density, but are not well suited for the evaluation of remodelling surfaces or dynamic parameters which have to be identified qualitatively before their quantification. In addition, in spite of recent advances in imaging techniques, these computers are not able to identify cells or staining and sectioning artefacts.

Finally, an efficient compromise is the use of a semi-automated image analyser (Malluche et al. 1981, 1982; Chavassieux et al. 1985a,b), which combines the advantages of discriminatory input by an investigator with reduced evaluation time resulting from computer analysis and calculations. The image of a magnetic table and of a luminous cursor is projected in the optical system of a microscope equipped with a drawing tube. The histomorphometrist chooses and traces all histological details to be quantified by moving the cursor on the table. This saves time and improves precision.

The following parameters are used to delineate the static and dynamic histomorphometric profiles which characterize the various osteopathies and will be defined. They give data not only on the amount of bone (calcified or not), but also on the activity of bone cells. Some are measured directly and others are calculated from the former. The new nomenclature for bone histomorphometry proposed by the ASBMR committee (Parfitt et al. 1987) is now widely used.

Structural static parameters measured in cortical bone are:

1. The cortical width (Ct-Wi) expressed in μm, which is the mean width of cortical bone tissue from the periosteal to the endosteal surfaces of cortices
2. The cortical porosity (Ct-Po), which is the percentage of a given volume of cortical bone tissue occupied by canals and resorption spaces. Only samples cut perpendicularly to the Haversian canals are measured.

Structural static parameters measured in cancellous bone are:

1. The cancellous bone volume (Cn-BV/TV), which represents the percentage of a given cancellous bone tissue occupied by trabeculae, excluding the medullary space, but including the calcified and the osteoid tissues
2. The cancellous mineralized volume (Cn-Md.V/TV), which is only the mineralized part of cancellous bone volume
3. The cancellous wall width (Cn-W.Wi), which is measured on totally mineralized bone packets without osteoid tissue and is expressed in μm.

Remodelling static parameters measured in cancellous bone are:

1. The cancellous eroded perimeter (Cn-E.Pm/B.Pm), the percentage of cancellous bone perimeter eroded, i.e. where osteoclastic

resorption is continuing or has ceased but where the osteoblasts have not yet started to refill Howship's lacunae

2. The cancellous osteoid volume (Cn-OV/BV), which represents the percentage of a given volume of cancellous bone occupied by osteoid tissue
3. The cancellous osteoid perimeter (Cn-O.Pm/B.Pm), which is the percentage of cancellous bone perimeter covered with osteoid tissue
4. The cancellous osteoid width (Cn-O.Wi), which is expressed in μm.

Cellular parameters measured in cancellous bone are:

1. The number of osteoclasts (Cn-N.Oc/B.Pm), which is counted per unit of cancellous perimeter
2. The cancellous osteoblast perimeter, which represents the percentage of bone perimeter covered with osteoblasts (Cn-Ob.Pm /B.Pm). Separate measurements between plump (Cn-plump Ob.Pm /B.Pm) or flat (Cn-flat Ob.Pm/B.Pm) osteoblasts may be made. These parameters may also be expressed as a percentage of cancellous osteoid perimeter (/O.Pm).

Dynamic parameters measured in cancellous bone of a patient given a double-labelling treatment include:

1. The cancellous mineral apposition rate (Cn-MAR), expressed in mm/day, which is the distance between two consecutive labels divided by the time between labelling periods
2. The cancellous single (Cn-sL.Pm/B.Pm) and double (Cn-dL.Pm /B.Pm) labelled perimeters which are the percentage of bone perimeter covered by a single or a double tetracycline label. The total labelled perimeter (Cn-L.Pm/B.Pm) is defined as the sum of sL.Pm + dL.Pm

3. The bone formation rate (Cn-BFR/B.Pm) gives the amount of mineralized bone formed per unit of bone perimeter per day and is expressed in $\mu m^2/\mu m$/day. This parameter is calculated as the product of Cn-MAR and labelled perimeters
4. The adjusted apposition rate (Cn-Aj.AR), which represents the amount of mineralized tissue being made per unit of osteoid-covered perimeter per day. This parameter is calculated as Cn-BFR/B.Pm divided by Cn-O.Pm/B.Pm and is expressed in μm/day
5. The cancellous mineralization lag time (Cn-Mlt) is the mean time interval between deposition and subsequent calcification of osteoid tissue. This parameter is calculated as Cn-O.Wi divided by Cn-Aj.AR and is expressed in days
6. The cancellous formation period (Cn-FP), expressed in days, represents the mean time required to rebuild a new bone packet and is calculated as Cn- W.Wi divided by Cn-Aj.AR. The active formation period (Cn-FP(a+)) is given as the ratio Cn-W.Wi divided by Cn-MAR
7. The activation frequency (Cn-Ac.f) is the probability that a new bone remodelling cycle will be initiated at any point on the surface of cancellous bone. It is calculated as Cn-BFR/B.Pm divided by Cn-W.Wi, and is expressed in $days^{-1}$.

Intra- and inter-observer coefficients of variation are generally less than 6 and 9% respectively if observers are experienced, and are higher in the manual method than in automatic and semi-automatic methods. Inter-method variations (Chavassieux et al. 1985a) are not very high but a given method appears better than others for a given parameter. The most sensitive method with the best reproducibility is the best one for the measurement of one given parameter. Thus, in routine practice, cancellous bone volume will be measured by an automatic method, osteoid volume by the manual one, and all other parameters by a semi-automatic method. Finally, automatic and semi-automatic methods seem to be the best ones, being more sensitive and less time-consuming. Inter-sample variations estimated on two contiguous iliac bone samples differ considerably between normal and pathologic subjects and from one parameter to another. For example, to be significant, in a single patient with osteoporosis, cancellous bone volume value has to be 30% different from the initial value. This is also the case for cancellous eroded perimeters in hyperparathyroidism (Chavassieux et al. 1985b).

5 An Example : Histomorphometric Findings in Skeletal Fluorosis

In the body, the main retention sites for fluoride ions are calcified tissues, mainly bone. The effects of fluoride on bone tissue depend on the quantity of fluoride ingested, on the time of exposure, and on the bioavailability of the fluoride salt used (Boivin and Meunier 1990). At very low doses (up to 1.5 mg F/day), fluoride prevents the development of tooth caries. At higher doses (about 23 mg F/day) it has a therapeutic action and is used in the curative treatment of type I osteoporosis with vertebral crush fractures. At doses higher than those mentioned and over a period of several years, fluoride may have toxic effects, causing the development of skeletal fluorosis of various aetiological origins (Boivin and Meunier 1990). Skeletal fluorosis is a bone disease resulting in a radiologically demonstrable abnormal bone densification. Bone fluoride content is very high (Boivin et al. 1988). Histological observations of bone tissue show osteosclerosis, mottled bone, formation defects and often hyperosteoidosis (Boivin et al. 1989, 1990). To date, very few studies have reported histomorphometric data compared with data on bone fluoride content in skeletal fluorosis. Thus, the main purpose of our study was to further investigate qualitative and quantitative bone modifications associated with skeletal fluorosis and, more precisely, the influence of fluoride on osteoblast number and function (Boivin et al. 1989 & 1990).

Histomorphometric analysis of undecalcified sections was performed in transiliac biopsy cores taken from 29 patients (16 men and 13 women aged 51 ±7 years) suffering from skeletal fluorosis as a result of chronic exposure to fluoride (Boivin et al. 1986, 1989, 1990). The origin of the exposure, known in 20 patients, was either hydric (endemic or sporadic), industrial or in a few cases, iatrogenic. Measured on calcined bone using a specific ion electrode (Boivin et al. 1988), bone fluoride content was significantly high in each specimen (mean ± SD : 0.79 ± 0.36% of bone ash) as compared to control values (less than 0.10%). The radiologically evident osteosclerosis observed in each patient was confirmed histologically (Fig. 3) by the significant increase of cancellous bone volume (40.1 ± 11.2% vs 19.0 ± 2.8% in controls, $p < 0.0001$). Typical histological modifications such as linear formation defects and mottled bone tissue, were observed in each patient (Fig. 4). There were significant increases in cortical width (1292 ± 395 μm vs 934 ±173 μm, $p < 0.0001$) and porosity (14.4 ± 6.4% vs 6.5 ± 1.7%, $p < 0.002$), but without reduction of cortical bone mass. Osteoid parameters were significantly increased in fluorotic patients. The increase of cancellous osteoid perimeter was almost three-fold greater than that noted in cancellous eroded perimeter. The fluorotic group had a greater number of osteoblasts than controls, with a very high proportion of flat osteoblasts. In 15 patients double-labelled with tetracycline, the mineral apposition rate was significantly decreased, while mineralization lag time significantly increased. Bone formation rate and adjusted apposition rate were significantly decreased in skeletal fluorosis. Cancellous wall width was normal in fluorosis but the formation period and active formation period were both significantly increased. Skeletal fluorosis is thus characterized by an unbalanced coupling in favour of bone formation, and also by a great number of osteoblasts with a high proportion of flat osteoblasts. The osteoblasts have a much longer formation period mainly due to a prolonged inactive formation period.

These findings support the view that fluoride may have a dual effect on osteoblasts; namely, an increased birth rate at the tissue level, and a toxic effect at the individual cell level. The addition of these two effects represents, however, a marked increase of bone formation at the organ level.

6 Other Fields of Application

In patients with all the typical clinical, biochemical and radiological signs of a well-defined metabolic bone disease, a bone biopsy for histomorphometry is not absolutely necessary for the diagnosis. However, in borderline cases the bone biopsy is extremely useful and can provide an unequivocal assessment of the histological diagnosis (Meunier 1981, 1983; Eriksen 1986).

Osteoporosis can be anatomically defined as an absolute decrease in the amount of bone tissue to an extent such that fractures (vertebral collapses or long bone fractures) may occur after minimal trauma. This level is defined as the "fracture threshold". Thus, the first information needed is an estimate of the amount of bone in the patient's body. To-date, it is not necessary to use a bone biopsy to obtain this information but non-invasive methods are well adapted to this (Meunier 1988). However, bone histomorphometry is the only method giving a direct and precise analysis of the static and dynamic cellular and tissue abnormalities responsible for bone loss (Meunier 1981, 1983, 1988; Arlot et al. 1990; Eriksen et al. 1990; Podenphant

1990). In osteoporosis, the decreased cancellous bone volume could result at the tissue level from decreased bone formation, increased bone resorption, or a combination of both (Riggs 1984; Arlot et al. 1990). In a recent study of the histological heterogeneity of osteoporosis, Arlot et al. (1990) have shown that the bone loss resulted from a wide spectrum of bone turnover abnormalities, with two distinct subsets: one with normal turnover and one with high turnover. Bone histomorphometry allowing a good approach at the intermediary level of bone organization is not only very useful as a diagnostic element, but also appears pertinent to the selection of the best treatment (Meunier 1988; Boivin and Meunier, 1990).

Histologically, osteomalacia is defined by the presence of both abnormally thick osteoid seams and significantly reduced calcification rate. An increase in the extent of osteoid seams only does not constitute an osteomalacia but reflects an increase in the number of apposition sites. This can be named "hyperosteoidosis" and it may be found in primary hyperparathyroidism, thyrotoxicosis and Paget's disease of bone (Meunier et al. 1977; Meunier 1981, 1983; Eriksen 1986).

The predominant histologic patterns of metabolic bone disease in uraemia are those of osteitis fibrosa and osteomalacia (Ritz et al. 1987). Renal osteodystrophy is a heterogeneous bone disorder which includes degrees of resorption, matrix formation and mineralization. The bone histomorphometric profile in uraemic patients depends mainly on two factors, parathyroid hormone secretion and the aluminum load. Parathyroid hormone is a potent activator of bone remodelling increasing both resorption and formation (Charhon et al. 1985a). Aluminum is often associated with low bone formation with or without osteomalacia (Charhon et al. 1985b, 1986). On the other hand, bone changes have been described at the time of, and following, renal transplantation (Charhon et al. 1983).

Paget's disease of bone is a benign bone disorder whose main features are hypertrophy and osteosclerosis of the skeletal parts. At the histological level, this bone disease is characterized by increased bone remodelling and an abnormally positive bone tissue balance (Meunier et al. 1980, 1987). Both hyperosteoclastosis and hyper-osteoblastosis seem to depend on a marked increase in the birth-rate of the basic multicellular units which remodel bone tissue. Pagetic bone not only reflects a disease but also represents an ideal model of bone hyper-remodelling useful for rapidly determining, by dynamic histomorphometry, the effects of any therapeutic agent on bone tissue and bone cells.

7 Summary

Adequate bone biopsy techniques providing a well-preserved sample and use of methods for preparing undecalcified bone specimens after tetracycline double labelling are prerequisites for quantitative evaluation of bone disease. Various methods are available for analysis of specimens, ranging from inexpensive but time-consuming manual procedures to semi-automatic or fully automatic methods with image-analysing computers.

Bone histomorphometry is a valid tool for diagnosing metabolic bone diseases. Endocrine osteopathies may also be identified by histomorphometry, although their histomorphometric profiles are not specific for given endocrine diseases and must be interpreted in the light of clinical and biochemical findings.

For research purposes longitudinal or cross-sectional studies of bone biopsy specimens in groups of patients are the only way to obtain direct patho-physiological information on the changes in the remodelling system that alter bone mass or bone mineral content.

Such analyses are, therefore, also of major importance in evaluating therapeutic effects (Melsen and Mosekilde, 1981). Furthermore, dynamic bone histomorphometry is the most appropriate available tool for studying the intermediary level of bone organization (the basic multicellular units and the basic structural units) which cannot be analysed through measurements of either bone mass (non-invasive methods) or biochemical markers in serum (osteocalcin) or urine (pyridinoline).

Acknowledgements. The authors thank Dr. P. Chavassieux for suggestions and criticisms during the preparation of this manuscript, M.J. Carew for reviewing the English text, and C. Fourneret for typing.

References

Arlot ME, Delmas PD, Chappard D, Meunier PJ (1990) Trabecular and endo-cortical bone remodeling in postmenopausal osteoporosis: comparison with normal postmenopausal women. Osteoporosis Int 1:41-49

Boivin G, Meunier PJ (1990) Fluoride and bone: toxicological and therapeutic aspects. In: Cohen RD, Lewis B, Alberti KGMM, Denman AM (eds) The metabolic and molecular basis of acquired disease, Baillire Tindall, London, pp 1803-1823

Boivin G, Chavassieux P, Chapuy MC, Baud CA, Meunier PJ (1986) Profil histomorphométrique de la fluorose osseuse induite par l'ingestion prolongeé d'eau de Vichy Saint-Yorre. Comparaison avec le taux de fluor osseux. Pathol Biol 34:33-39

Boivin G, Chapuy MC, Baud CA, Meunier PJ (1988) Fluoride content in human iliac bone: results in controls, patients with fluorosis, and osteoporotic patients treated with fluoride. J Bone Mineral Res 3:497-502

Boivin G, Chavassieux P, Chapuy MC, Baud CA, Meunier PJ (1989) Skeletal fluorosis: histomorphometric analysis of bone changes and bone fluoride content in 29 patients. Bone 10:89-99

Boivin G, Chavassieux P, Chapuy MC, Baud CA, Meunier PJ (1990) Skeletal fluorosis: histomorphometric findings. J Bone Mineral Res 5 Suppl 1:185-189

Bordier P, Matrajt H, Miravet L, Hioco D (1964) Mesure histologique de la masse et de la résorption des traves osseuses. Pathol Biol 12:1238-1243

Charhon SA, El Yafi S, Oko A, Edouard C, Traeger J, Meunier PJ (1983) Bone changes following renal transplantation. In: Touraine JL, Traeger J, Btuel H, Brochier J, Dubernard JM, Revillard JP, Triau R (eds) Transplantation and clinical immunology, vol 15, Excerpta Medica, Amsterdam, pp 186-196

Charhon SA, Berland YF, Olmer MJ, Delawari E, Traeger J, Meunier PJ (1985a) Effects of parathyroidectomy on bone formation and mineralization in hemodialyzed patients. Kidney Int 27:426-435

Charhon SA, Chavassieux PM, Chapuy MC, Boivin GY, Meunier PJ (1985b) Low rate of bone formation with or without histologic appearance of osteomalacia in patients with aluminum intoxication. J Lab Clin Med 106:123-131

Charhon SA, Delmas PD, Malaval L, Chavassieux PM, Arlot M, Chapuy MC, Meunier PJ (1986) Serum bone Gla-protein in renal osteodystrophy: comparison with bone histomorphometry. J Clin Endocrinol Metab 63:892-897

Chavassieux PM, Arlot ME, Meunier PJ (1985a) Intermethod variation in bone histomorphometry: comparison between manual and computerized methods applied to iliac bone biopsies. Bone 6:221-229

Chavassieux PM, Arlot ME, Meunier PJ (1985b) Intersample variation in bone histomorphometry: comparison between parameter values measured on two contiguous transiliac bone biopsies. Calcif Tissue Int 37:345-350

Delesse M (1847) Procédé mécanique pour déterminer la composition des roches. C R Acad Sci Paris 25:544-552

Dickson GR (ed) (1984) Methods of calcified tissue preparation. Elsevier, Amsterdam

Duncan H, Rao SD, Parfitt AM (1981) Complications of bone biopsies. In : Jee WSS, Parfitt AM (eds) Bone histomorphometry 1980. Société Nouvelle de Publications Médicales et Dentaires, Paris, pp 483-486

Epker BN, Frost HM (1965) A histological study of remodeling at the periosteal, haversian canal, cortical endosteal and trabecular endosteal surfaces in human ribs. Anat Rec 152:129-135

Eriksen EF (1986) Normal and pathological remodeling of human trabecular bone: three dimensional reconstruction of the remodeling sequence in normals and in metabolic bone disease. Endocrine Rev 7:379-408

Eriksen EF, Hodgson SF, Eastell R, Cedel SL, O'Fallon WM, Riggs BL (1990) Cancellous bone remodeling in type I (postmenopausal) osteoporosis: quantitative assessment of rates of formation, resorption, and bone loss at tissue and cellular levels. J Bone Mineral Res 5:311-319

Frost HM (1969) Tetracycline based histological analysis of bone remodeling. Calcif Tissue Res 3:211-237

Frost HM (1973) Bone remodeling and its relation to metabolic bone disease. Thomas, Springfield, Illinois

Frost HM (1983) The skeletal intermediary organization. Metab Bone Disease Rel Res 4:281-290

Jaworski ZFG (ed) (1976) Proc 1st Int Worksh Bone histomorphometry, Ottawa 1973. Univ Press, Ottawa

Jaworski ZFG (1982) Bone histomorphometry: outline of theory and practice. In: Skeletal research, vol 2. Academic Press, NewYork, London pp 237-276

Jee WSS, Parfitt AM (eds) (1981) Proc 3rd Int Worksh Bone histomorphometry, Sun Valley 1980. Société Nouvelle de Publications Médicales et Dentaires, Paris

Jowsey J (1977) The bone biopsy. Plenum, New-York

Jowsey J, Kelly PJ, Riggs BL, Bianco AJ, Scholz DA, Gershon-Cohen J (1965) Quantitative microradiographic studies of normal and osteoporotic bone. J Bone Joint Surg 47A:785-806

Malluche HH, Sherman D, Manaka R, Massry SG (1981) Comparison between different histomorphometric methods. In: Jee WSS, Parfitt AM (eds) Bone histomorphometry 1980. Société Nouvelle de Publications Médicales et Dentaires, Paris, pp 449-451

Malluche HH, Sherman D, Meyer W, Massry SG (1982) A new semiautomatic method for quantitative static and dynamic bone histology. Calcif Tissue Int 34:439-448

Melsen F, Mosekilde L (1981) The role of bone biopsy in the diagnosis of metabolic bone disease. Orthop Clin N Am 12:571-602

Meunier PJ (1973) Use of an image-analyzing computer for bone morphometry. In: Frame B, Parfitt AM, Duncan H (eds) Clinical aspects of metabolic bone disease. Excerpta Medica, Amsterdam, pp 148-151

Meunier PJ (1977) Proc 2nd Int Worksh Bone histomorphometry, Lyon 1976. Société de la Nouvelle Imprimerie Fournié, Toulouse

Meunier PJ (1981) Bone biopsy in diagnosis of metabolic bone disease. In: Cohn DV, Talmage RV, Matthews JL (eds) Hormonal control of calcium metabolism, Excerpta Medica, Amsterdam, pp 109-117

Meunier PJ (1983) Histomorphometry of the skeleton. In: Peck WA (ed) Bone and mineral research, Annual 1. Elsevier, Amsterdam, pp 191-222

Meunier PJ (1988) Assessment of bone turnover by histomorphometry in osteoporosis. In: Riggs BL, Melton LJ III (eds) Osteoporosis: etiology, diagnosis, and management, Raven, New York, pp 317-332

Meunier PJ, Edouard C, Courpron P, Toussaint F (1975) Morphometric analysis of osteoid in iliac trabecular bone. In: Norman AW, Schaefer K, Grigoleit HG, von Herath D, Ritz E (eds) Vitamin D and problems related to uremic bone disease. De Gruyter, Berlin, pp 149-155

Meunier PJ, Edouard C, Richard D, Laurent J (1977) Histomorphometry of osteoid tissue. The hyperosteoidoses. In: Meunier PJ (ed) Proc 2nd Int Worksh Bone histomorphometry. Société de la Nouvelle Imprimerie Fournié, Toulouse, pp 249-262

Meunier PJ, Coindre JM, Edouard CM, Arlot ME (1980) Bone histomorphometry in Paget's disease. Quantitative and dynamic analysis of pagetic and nonpagetic bone tissue. Arthritis Rheumat 23:1095-1103

Meunier PJ, Salson C, Mathieu L, Chapuy MC, Delmas PD, Alexandre C, Charhon SA (1987) Skeletal distribution and biochemical parameters of Paget's disease. Clin Orthop 217:37-44

Parfitt AM, Drezner MK, Glorieux FH, Kanis JA, Malluche H, Meunier PJ, Ott SM, Recker RR (1987) Bone histomorphometry: standardization of nomenclature, symbols, and units. J Bone Mineral Res 2:595-610

Podenphant J (1990) Methodological problems in bone histomorphometry and it's application in postmenopausal osteoporosis. Dan Med Bull 37:424-433

Riggs BL (1984) Treatment of osteoporosis with sodium fluoride: an appraisal. In: Peck WA (ed) Bone and Mineral Research, Annual 2. Elsevier, Amsterdam, pp 366-393

Ritz E, Dreke T, Merke J, Lucas PA (1987) Genesis of bone disease in uremia. In: Peck WA (ed) Bone and Mineral Research, Annual 5. Elsevier, Amsterdam, pp 309-374

Sacker LS, Nordin BEC (1954) A sample bone biopsy needle. Lancet 1:347

Schenk RK, Merz WA, Müller JA (1969) A quantitative histological study of bone resorption in human cancellous bone. Acta Anat 74:44-53

Schenk RK, Olah AJ, Herrmann W (1984) Preparation of calcified tissues for light microscopy. In: Dickson GR (ed) Methods of calcified tissue preparation. Elsevier, Amsterdam, pp 1-56

Palaeohistology of Human Bone Remains: a Critical Evaluation and an Example of Its Use

Paolo Bianco and Antonio Ascenzi
Dipartimento di Biopathologia Umana
Sezione di Anatomia Patologica
Università "La Sapienza"
Viale Regina Elena 324
I-00161 Roma
Italy

1 The Value and Limitations of Histology in the Pathological Assessment of Archaeological Human Skeletal Remains

The-ever growing interest of archaeologists, palaeontologists and palaeopathologists in the techniques of microscopy undoubtedly stems from a host of solid contributions these techniques have made in helping to understand the nature and mechanism of the processes of fossilization, as well as of a variety of diagenetic changes that buried bones undergo. A survey of these contributions is not in order here as it is available elsewhere (Ascenzi 1969, 1986). Rather, this paper will focus on the issue of merit which histology (or more broadly, microscopy) has as an asset to the palaeopathological evaluation of archaeological specimens. Dealing with this topic requires a somewhat critical attitude, which would reduce the risk of hurrying over, or the understanding of the value and meaning of palaeohistology. Also, bone, as opposed to mummified tissues, poses special problems in the evaluation of pathological changes. It should not be forgotten that bone palaeopathology has mostly relied on gross and X-ray evaluation, with histology playing a minor role so far, whereas histopathological studies of mummies have led in many instances to significant contributions. This stems from the fact that archaeological bone is a cell-free and soft tissue free bone.

Bone reacts to disease in a limited number of ways. Only in a restricted subset of pathological conditions do basic tissue reactions produce changes in the mineralized matrix apt to be recognized as specific hallmarks of specific disease processes. The areas of overlap of observable changes in the mineralized matrix are plenty and broad, and it is only a set of changes in the bone cell population, the amount and distribution of unmineralized bone matrix, and the reaction of adjacent soft tissues which makes discrimination between different diseases affordable in present day bone pathology. The palaeopathologist looking at a bone specimen is thus deprived of most key diagnostic features available to present-day pathologists. In addition, clinical data, patient follow-up and databases are not available in palaeopathology as an independent source of information to validate (or "falsify") estimated diagnoses. The reason why palaeopathologists should be aware of these limitations is the need to recognize that perhaps the main risk they face is that of making non-scientific statements (statements which cannot be proven false) rather than 'false' statements, which is the daily risk of every practising pathologist.

A host of microscope technologies are variably suited to studying archaeological bone. Polarized light microscopy, microradiography, electron microscopy (both transmission and scanning) and more recently confocal microscopy can be profitably applied to the study of ancient bones. Nonetheless, the nature of archaeological bone narrows the range of changes that can be detected. These include the detection of woven as opposed to lamellar bone, assessment of number, size, density and orientation of osteocyte lacunae (an additional key feature in the discrimination of woven versus lamellar bone), osteosclerosis (increased density of bone trabeculae, i.e. increased bone mass per unit volume), rarefaction of bone (decreased bone mass per unit volume), extent of the process of bone remodelling as assessed by microradiography, occurrence of special internal features of bone trabeculae (such as the time-honoured Schmorl's mosaic), and the detection of abnormally sized and shaped collagen fibres in bone (provided the process of fossilization has not progressed so far as to efface their individuality by the electron microscope). No doubt there are a few bona fide diagnoses which can be made on the basis of this type of information. For example, osteogenesis imperfecta ossium and fibrogenesis imperfecta ossium, two inherited diseases resulting in abnormal assembly of collagen fibres, are, in principle, suited to recognition by electron microscopy in ancient specimens, provided of course that the process of fossilization has not progressed so far as to efface their individuality. Also, Schmorl's mosaic can be recognized in ancient bones; however, it should be borne in mind that its specificity for Paget's disease is perhaps far less than absolute, if not matched to other features, none of which are, in turn, specific (i.e. gross changes, site of the lesions, detection of virus-like particles in osteoclast nuclei, changes in marrow vascularity and stromal architecture), but many of which cannot be evaluated in ancient specimens. On the other hand, the detection of a rarefaction of bone trabeculae with a reduction in the mineralized bone mass can easily be detected in an archaeological bone studied by microscopy and may, on most occasions, not be equated with a "diagnosis". In fact, it can be the final common pathway for many different bone diseases which only differ from one another with respect to the changes in bone cell numbers and activities. Perhaps the best known example to present day bone pathologists is the discrimination between osteoporosis, osteomalacia, and hyperparathyroidism-related osteopenia. The impact that recognition of some of these diseases has on the evaluation of nutrition and ageing of individuals, as well as of populations, cannot be overlooked, and make the indication of such potential areas of uncertainty something more than an ad hoc argument for cautious use of palaeohistological data.

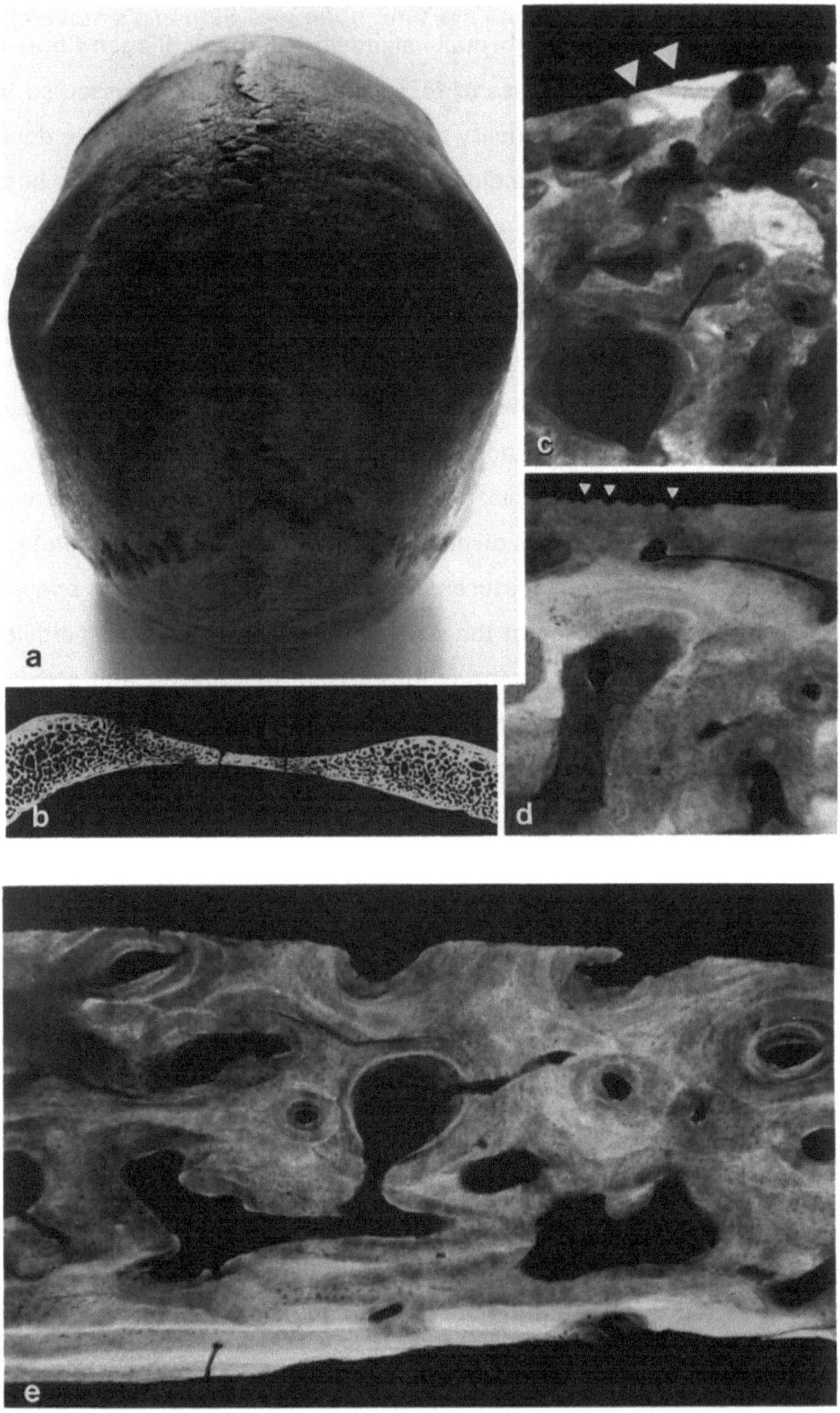
a
b
c
d
e

Fig. 1

a Posterior view of the ancient Egyptian skull. The two fairly rounded and fairly symmetrical zones of thinness of the parietal bones are clearly visible.

b Low power view of the microradiography of a 100 μm thick section obtained from a slice of thinned parietal bone. The non-thinned edges of the lesions are included in the specimen. Note the presence of distinct inner and outer layers of compact bone encasing the trabecular bone of diploe in the non-thinned areas. Also note that no diploe is left at the site of maximal thinning, where the bone appears compact.

c Detail of the microradiograph showing the transition from the normal bone to the thinned area. The profile of bone cuts at an angle of about 40° the direction of the lamellae of the outer system, which appears to be sharply truncated (arrow). Note the abundance of osteons at low grade of calcification, indicating active bone apposition deep to the surface of bone undergoing 'thinning'.

d Detail of the bone surface in the immediate vicinity of the thinned zone. Note the scalloped profile of the surface due to several Howship's lacunae (sites of osteoclastic bone resorption, arrows), and the apposition of new osteonal bone deep to the surface.

e High-power view of the zone of maximal thinning. The whole field is taken up by compact, mostly osteonal bone, with no remnants of diploic trabecular bone. The continuous lamellar structure of the inner table is preserved and visible at the bottom, whereas its outermost counterpart has disappeared, and the outer surface shows "opened up" Haversian systems

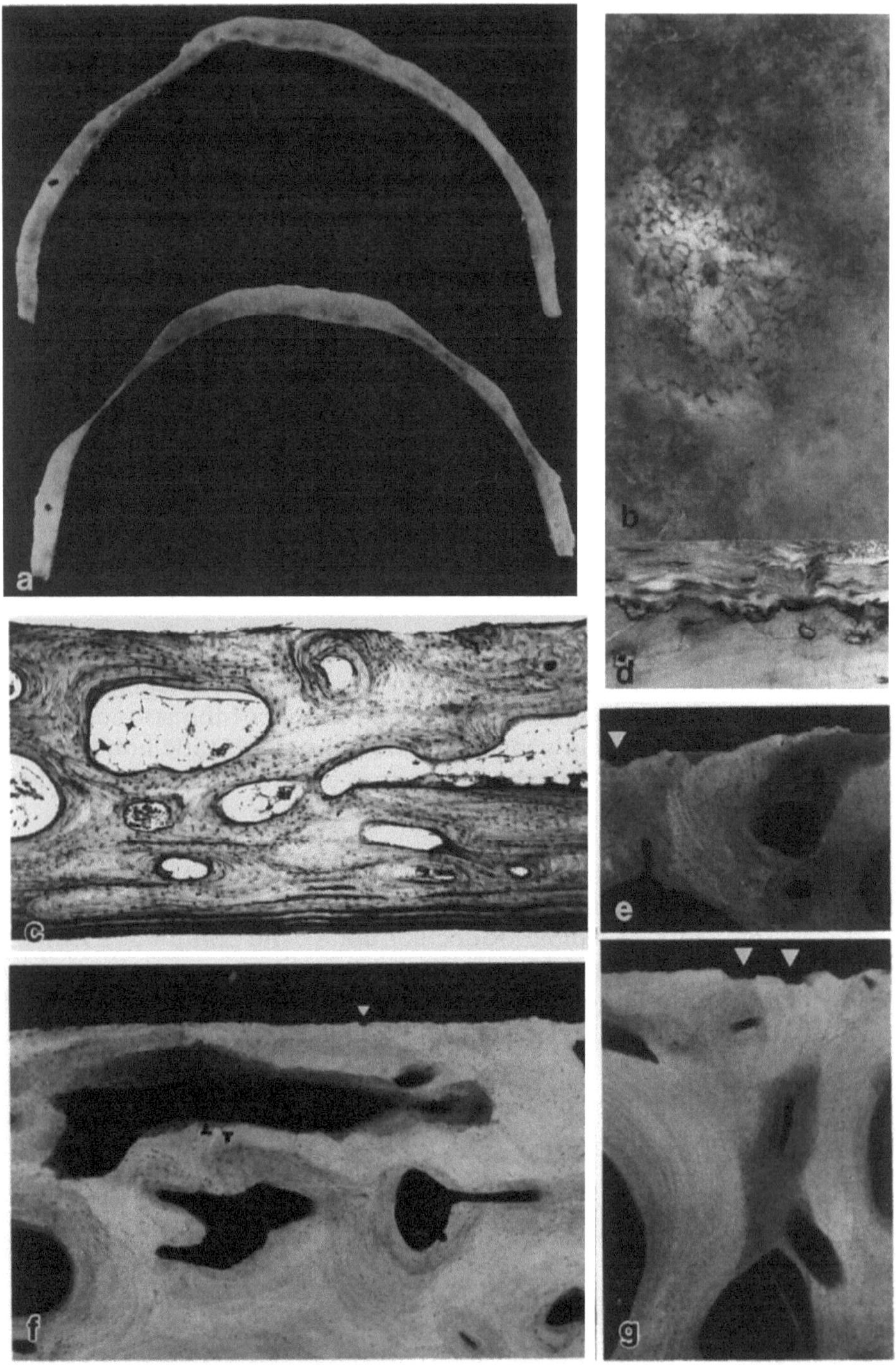
a
b
c
d
e
f
g

Fig. 2

a Frontal sections of a present day case of 'biparietal thinning'. Because of lack of perfect symmetry of the zones of thinning, the sections do not cut across both sites of maximal thinning.

b Direct view of one thinned area in the transilluminated, intact skull. Note the vascular pattern of diploic vessels clearly depicted in the thinned zone.

c Histological section of periosteum-stripped zone of maximal thinning. Note the architectural resemblance with the zone of maximal thinning of the ancient specimen shown in Fig. 1e.

d Histological section of the zone of thinning; the remaining periosteum is at variance with the situation in **c**. High-power view showing the scalloped profile of the periosteal bone surface due to the presence of many Howship's lacunae, mostly occupied by mononuclear cells.

e-g Details of microradiographs showing resorption lacunae at the surface of bone (arrows) and several osteons at low grade of calcification, indicating the simultaneous occurrence of bone resorption at the surface, and formation in the underlying tissue

A further important point is the two-way interplay of pathological and diagenetic changes. Palaeohistology has been instrumental in determining the influence that natural bone architecture has on the non-random progression and spread of diagenetic changes, which implies that pathology (which results in altered architecture) also has a bearing on diagenetic changes. How diagenesis can result in effacement of pathological changes or in mimicing such changes has also been recently stressed (Bell 1990).

A common approach to understanding the origin and nature of diagenetic microscopic changes as detected in buried bone has entailed the use of an empirical reproduction of the changes in modern day specimens under controlled conditions. The role of microorganisms, soil and sea water in generating diagenetic changes in the structure of bones has been investigated using this approach (Marchiafava et al. 1974; Arnaud et al. 1978; Ascenzi and Silvestrini 1984). A similar approach might be of value, at least in a few disease states, when applied to changes that are pathological rather than taphonomical in nature. This approach would entail the study of modern day specimens under conditions as close as possible to those encountered in the study of ancient specimens. Bone pathologists with an interest in palaeopathology should become familiar with the features exhibited by the specimens they observe in their fresh state when examined by the more common and feasible approaches of palaeohistology. They should start to compare histology of fresh bones with their cells and associated soft tissues with microradiographic features and, when available with inorganic preparations studied in the SEM. This will, in time, help with the construction of a database for palaeohistopathology of bone, as well as with identification of subtle changes of diagnostic significance which could be evaluated and detected in cell-free, soft tissue free bone.

The following pages will be devoted to a 'sample study' illustrating the potential value of the combined use of 'palaeo' and 'neo' histopathology of bone.

2 A Palaeohistopathology Case: Biparietal Atrophy in an Egyptian Dynastic Skull and Comparison with its Present-Day Counterpart

Because of its unusually high prevalence in ancient Egypt, as disclosed by the classic studies of Elliot Smith (1907), "bilateral thinning of parietal bones" is a prime issue in palaeopathology of bone (Lodge 1967). The lesion has been traced back to as early as the Bronze Age (Dutta 1969), and has been known to the medical world since the time of Virchow but its aetiology and nature has remained elusive. Virchow (1854) considered the lesion as a localized form of senile osteoporosis, a view also shared by other 19th century pathologists, like Rokitansky. This view has now been abandoned since Camp and Nash (1944) reported the study of a large series of present-day cases which included young adults and children. The alternative hypothesis put forward by Elliot Smith (1907) was the consideration of the lesion as a localized pressure atrophy from wearing heavy wigs but this has also been challenged. Although seemingly uncommon, the lesion is not exceptional in present day pathology. Because of its almost constant lack of symptoms, it is, as a rule, discovered as a chance finding upon autopsy or on the occasion of X-ray examination of the skull for unrelated purposes. Virchow's and Lobstein's classical histological studies of this intriguing condition date back to the 19th century, whereas no documentation of its histology is available in recent literature, a fact which can perhaps be seen as one for the reasons of the poor understanding of its genesis.

An Egyptian skull of the dynastic period, exhibiting typical biparietal thinning and conserved at the Egyptian Museum of Turin, Italy, and a present-day case were studied histologically in our laboratory with the aim of comparing the 'cell-free' bone histology in ancient and present-day specimens, and also of evaluating the participation of cellular and soft tissue components in the pathogenesis of the lesion.

Figure 1 shows the macroscopic appearance of the lesions as seen in the Egyptian dynastic skull. Two fairly round and fairly symmetrical depressions occupy the outer aspect of the two parietal bones between the sagittal suture and the upper temporal line, a gross morphology which overlaps that of other known dynastic skulls, such as Khety, Meritamon, and Tutmosis III. A slice of bone running across the depressed zone, and including adjacent portions of parietal bone, was taken for microscopic examination, embedded in methylmethacrylate, sectioned to a thickness of 100 μm, and viewed by polarized light microscopy. Contact microradiographs were also prepared and studied.

A low power inspection of contact microradiographs (Fig. 1b) showed that at the site of maximal depression the bone thickness was greatly reduced, but the bone density was remarkably increased, as if the whole (reduced) thickness was taken up by compact bone. The depression affected the outer profile of bone, not the inner one. The normal architecture was well preserved in the non-thinned zones of parietal bone, in which the outer table, the diploe and the inner table could be easily seen even at low power magnification (Fig. 1b). Further detail was demonstrated at higher power: (1) an outermost lamellar system made of mostly non-osteonal bone high in mineral content, (2) a zone of trabecular bone harbouring marrow spaces (the diploe); (3) a deeper zone of more compact bone, in which trabecular bone converts into osteonal compact bone; and finally (4) an inner lamellar system. At the edge of the thinned part the outer lamellar system was found to abruptly disappear (Fig. 1c). There was a progressive reduction in thickness of the trabecular diploic bone moving from the edge to the centre of the depressed zone. At the point of maximal thinness (Fig. 1e) no diploe was discernible any more. Here, the bone was largely osteonal in architecture, and the inner lamellar layer high in mineral content was easily discernible. Most of the outer surface of the thinned bone appeared scalloped with the presence of many Howship's lacunae, indicating active bone resorption. These lacunae could also be seen at the periosteal surface of the non-thinned bone immediately surrounding the depressed zone (Fig. 1d). There was no evidence of osteoporotic changes, nor did polarized microscopy demonstrate any primary (woven) bone.

Figure 2 shows for comparison frontal sections of a present-day skull with biparietal thinning and the appearance of one of the zones of "thinning" as seen by direct macroscopic inspection. Except for the lack of perfect symmetry of the two zones of thinning which a comparison of the frontal sections demonstrates (in agreement with the observations of others), the gross morphology does not differ from that of the ancient Egyptian skull, and the inner profile of the cranial vault is regular. In addition to ground sections and contact microradiographs, conventional 5-8 μm thick sections of demineralized, paraffin embedded samples were prepared from this specimen.

The microradiographs demonstrated a state of moderate osteoporosis, as indicated by the thinness of trabeculae and the increased distance between individual trabeculae resulting in increased width of marrow spaces. Most likely, because of this osteoporotic background, relatively broad diploic spaces could also be demonstrated at the site of maximal thinning, at variance with the ancient Egyptian sample. The abrupt disappearance of the outer lamellar system and indirect signs of bone resorption at the subperiosteal surface (Howship's lacunae) were, however, detected in the present-day specimens as well.

Conventional histology and transmission electron microscopy (not shown here) were instrumental in this case to detect osteoclasts actively engaged in subperiosteal resorption. Of note, most of them were mono- or binuclear, with classical Koelliker's multinuclear cells being remarkably few in number.

Regardless of the concomitance of osteoporotic changes, the present-day case and the palaeopathological specimen thus displayed the same basic changes. These changes allowed us to draw the conclusions that firstly, "biparietal thinning" of ancient Egyptian skulls is not a localized form of senile osteoporosis; secondly, the prime event is an outscale resorptive activity starting at the subperiosteal bone surface, involving the outer lamellar system first and then spreading inwards; and thirdly, in the absence of concomitant osteoporotic changes, the process of 'thinning' entails the disappearance of diploic trabeculae and marrow spaces, leading to a condition that is, in a way, opposite to the "porotic hyperostosis" (i.e. a "sclerosing atrophy"). The absence of concomitant changes in mineral metabolism, familiarity or cranial soft tissue and of evidence of post-cranial bone disease in the present-day case indicated the apparent primary character of the cranial lesions. What the combination of 'palaeo' and 'neo' histology demonstrated was the activation of unbalanced (uncoupled) bone resorption in a strictly localized fashion. The role of osteoclastic bone resorption in the pathogenesis of the lesion was postulated by Virchow (1854), but had never been demonstrated in ancient specimens before. The comparative study of ancient and present-day specimens strongly suggested that the nature of the lesion was the same in both circumstances, and also ruled out many of the hypotheses put forward at various times by different authors, such as the habit of wearing heavy wigs, or a primary dysplasia of the diploe.

The problem of the origin and nature of the lesion can thus be rephrased as: what is the prime biological event leading to a highly localized, unbalanced osteoclast activation? Answering this question will require further study on current models of the ancient disease which, if successful, would provide important information on the local factors involved in a natural "model system" of strictly localized bone resorption and will in turn lead to potentially important insights into the so far unexplained, unusually high prevalence of the condition in ancient Egypt.

Acknowledgements. We thank Dr Elena Rabino Massa of the Egyptian Museum, Turin, Italy, for allowing us to study the archaeological specimen described herein. We also thank Mr Alessandro Benvenuti for skilled technical help.

References

Arnaud G, Arnaud S, Ascenzi A, Bonucci E, Graziani G (1978) On the problem of the preservation of human bone in sea water. J Human Evol 7:409-420

Ascenzi A (1969) Microscopy and prehistoric bone. In: Brothwell D, Higgs E (eds) Science in archaeology. Thames & Hudson, London, pp 526-538

Ascenzi A (1986) Microscopy and ultramicroscopy in palaeopathology. In Herrmann B (ed) Innovative trends in prehistoric anthropology. Mitt Berliner Ges Anthropol Ethnol und Urgesch 7:149-156

Ascenzi A, Silvestrini G (1984) Bone-boring marine microorganisms: an experimental investigation. J Human Evol 13:531-536

Bell LS (1990) Palaeopathology and diagenesis: an SEM evaluation of structural changes using backscattered electron imaging. J Archaeol Sci 17:85-102

Camp JD, Nash LA (1944) Developmental thinness of the parietal bones. Radiology 42:42-47

Dutta PC (1969) Bilateral parietal thinning in Bronze Age skull. Brit Med J 1:55

Elliot Smith G (1907) The causation of the symmetrical thinning of the parietal bones in ancient Egyptians. J Anat Physiol London 41:232-241

Lodge T (1967) Thinning of the parietal bones in early Egyptian populations and its aetiology in the light of modern observations. In: Brothwell D, Sandison AT (eds) Disease in antiquity. Springfield, Illinois, pp 405-412

Marchiafava V, Bonucci E, Ascenzi A (1974) Fungal osteoclasia: a model of dead bone resorption. Calcif Tissue Res 14:195-210

Virchow R (1854) Über die Involutionskrankheit (Malum senile) der platten Knochen. Verhandl Phys Med Ges Würzburg 4:354

Pathology of Metabolic Bone and Joint Diseases

Brendan F. Boyce
Department of Pathology
The University of Texas
Health Science Center at San Antonio
7703 Floyd Curl Drive
San Antonio
Texas 78284-7750
USA

1 Introduction

Adult human bone is continually being remodelled by a process in which small microscopic packets are removed by osteoclasts from trabecular bone surfaces or from within compact bone (Parfitt 1986). They are replaced at these sites by new osteoid matrix which is laid down and calcified by osteoblasts. There are 1-2 million bone remodelling sites or units in the normal adult human skeleton and their number may be increased in conditions in which there is excessive local or systemic release of osteoclast-activating factors. Bone formation is typically coupled to sites of bone resorption and thus normal bone remodelling does not significantly disturb the bone microarchitecture. In many conditions, however, the local or systemic release of osteoclast or osteoblast stimulating factors can lead to osteopenia or to osteolytic or osteosclerotic lesions which may be visible radiologically. In this chapter, the pathological features of the main metabolic bone and joint diseases will be outlined with an emphasis on how excessive local or systemic secretion of osteotropic factors can lead to gross and/or microscopic abnormalities of bone architecture.

The main, known osteoclast activating factors are listed in Table 1. Some of these, particularly the endocrine hormones, stimulate osteoclastic resorption systemically while others, such as cytokines and growth factors, are more likely to act locally on bone surfaces. Present evidence suggests that all these factors exert their effects on osteoclasts indirectly through osteoblasts since these latter cells, rather than the former, express the appropriate cell surface receptors (Chambers 1985). Upon stimulation, osteoblasts release an osteoclast-stimulating factor(s) which causes the local activation of resorption.

TABLE 1. The main osteoclast-activating factors

Osteoclast stimulating factors	**Known diseases**
PTH,(PTH-rP), 1,25$(OH)_2$ Vitamin D_3 Oestrogen lack Corticosteroids Thyroid hormone	Hyperparathyroidism Osteoporosis Cushing's syndrome Thyrotoxicosis
IL-1, TNF, IL-6 Prostaglandins ODFR	Rheumatoid arthritis Inflammation in bone Periodontal disease
IL-1, TNF, LT, Prostaglandins TGFa, EGF, PDGF, GM-CSF, M-CSF, PTH-rP	Hypercalcaemia of malignancy Metastatic tumours in bone Multiple myeloma

Abbreviations: PTH, parathyroid hormone; PTH-rP, parathyroid hormone-related protein; IL-1, interleukin-1; TNF, tumour necrosis factor; IL-6, interleukin 6; ODFR, oxygen-derived free radicals; LT, lymphotoxin; TGFa, transforming growth factor a; EGF, epidermal growth factor; PDGF, platelet-derived growth factor; GM-CSF, granulocyte/monocyte colony stimulating factor;M-CSF, monocyte colony stimulating factor

2.1 Hyperparathyroidism

Primary hyperparathyroidism occurs most commonly in postmenopausal females. It is usually caused by an adenoma of one of the parathyroid glands (85%) or, less commonly, by primary hyperplasia of all four glands (15%). Less than 1% of cases are due to parathyroid carcinoma. The typical effects of parathyroid hormone (PTH) on bone are seen in Fig. 1 which shows mild to moderate osteitis fibrosa, the description given to the increased osteoclastic resorption, new bone formation and marrow fibrosis. Bone turnover, however, may be normal (10-20% of cases) or may be greatly increased. There is a tendency for resorption to be increased relative to formation and this may contribute to the development of osteoporosis in some patients. The increased turnover of hyperparathyroidism results in an increase in the number of bone remodelling units and in the irregularity of bone surfaces at the microscopic level (Fig. 1). This may be accompanied by gross irregularity of the outer surface of some bones, particularly the phalanges of the hand, that can be seen radiologically as subperiosteal bone resorption. In a small proportion of cases, gross cystic lesions may appear in long bones giving rise to the radiological appearance of osteitis fibrosa cystica in which the "cysts" are filled with fibrous tissue and many osteoclasts.

2.2 Renal Bone Disease

Hyperparathyroidism may be secondary to a low blood calcium concentration (hypocalcaemia) which may accompany chronic kidney diseases. The underlying defect is the kidney's failure to excrete phosphate. This leads to hyperphosphataemia with resulting hypocalcaemia, increased PTH secretion (secondary hyperparathyroidism), osteoclast activation and osteitis fibrosa. Since the kidney is the major site for the hydroxylation of 25(OH) vitamin D_3 to 1,25$(OH)_2$ vitamin D_3, the most active vitamin D metabolite, some patients may also develop osteomalacia due to reduced activity of the 1-hydroxylase enzyme in damaged renal tubular cells.

In patients with end-stage chronic renal failure who have not yet been treated with haemodialysis, 60-70% have osteitis fibrosa, and 20-30% have osteomalacia (Ellis and Pearl 1973). In each of these conditions, the appearance of the edges of calcified bone may be similarly irregular reflecting the excess of osteoclastic bone resorption. Thus, although mineralization is normal in patients with secondary hyperparathyroidism and defective in osteomalacia leading to extensive, thickened osteoid seams, the skeletal remains of both types of patient might appear very similar after the unmineralized osteoid has decomposed.

A further form of bone disease that may occur in patients with chronic renal failure is aluminium-related osteomalacia due to exposure to excessive amounts of aluminium. The two major sources of aluminium for these patients are the water used during dialysis and aluminium-containing phosphate binding agents which are taken orally by patients in an attempt to reduce absorption of phosphate from the gastrointestinal tract and thus prevent hyperphosphataemia. Some of this aluminium may be absorbed by the gut and lead to high levels of aluminium in the blood. Aluminium concentrations may be elevated in the dialysis water due to the addition of aluminium salts to some public water supplies in an attempt to clear water of algae and other agents that cause aesthetically unacceptable discolouration of drinking water.

The aluminium binds to mineralizing surfaces on bone and inhibits calcification and thus causes osteomalacia. Its localization at the mineralization front was first detected using the technique of electron probe X-ray microanalysis which can detect the site of localization of small concentrations of many elements within tissues (Boyce et al. 1985). Techniques such as this, ion emission spectroscopy, and laser microprobe mass analysis (Verbeuken et al. 1985) might be applicable to the analysis of ancient specimens of diseased bones providing the problems associated with the encrustation of human remains with elements (diagenesis) can be well characterized.

2.3 Vitamin D Deficiency-Related Osteomalacia

Vitamin D deficiency can result in hypocalcaemia due to defective absorption of calcium from the gastrointestinal tract. The resulting secondary hyperparathyroidism causes increased osteoclastic bone resorption and the formation of new bone matrix which does not calcify normally because vitamin D is required for normal mineralization. This is the commonest form of osteomalacia and is seen most frequently in individuals living in poor social conditions with low exposure to sunlight. These conditions prevailed in some parts of Europe following the Industrial Revolution with its associated atmospheric pollution. The young and the elderly are particularly at risk of vitamin D deficiency since, on the one hand children have increased requirements for calcium for bone growth and calcification and, on the other hand, elderly subjects are less active and tend to have minimal exposure to sunlight. Ultraviolet irradiation causes the production in the skin of cholecalciferol, a precursor of vitamin D.

Vitamin D deficiency in children may cause rickets which is characterized by defective mineralization of bone and cartilage. This can result in soft bones that bend under normal body weight causing bowing of the tibiae or femora. Subsequent mineralization of these deformed bones can result in the persistence of skeletal deformity into old age. By contrast, the development of vitamin D deficiency in old age and resulting secondary hyperparathyroidism and osteomalacia do not lead to the bowing of legs, as in children, but to a reduction in the volume of calcified matrix and an increase in unmineralized osteoid. The microscopic appearance of the skeletal remains from such patients might reveal irregularity of the calcified bone surface, due to increased osteoclastic resorption, and a reduction in calcified bone mass. Despite the reduction in calcified bone volume, the appearances would differ from those of age-related or postmenopausal osteoporosis since in both of these conditions, although there is excessive osteoclastic bone resorption relative to the amount of new bone formation, bone surfaces tend to be smooth.

2.4 Osteoporosis

Osteoporosis occurs in males and females as a result of an age-related reduction in the amount of matrix laid down by osteoblasts relative to the amount of bone removed by osteoclasts during bone remodelling. This can lead to a reduction in cortical bone thickness and in the mean thickness of trabeculae with an increased risk of fracture. Furthermore, resorption by osteoclasts through the full thickness of trabecular plates can account for the appearance of isolated trabecular elements in histological sections of bone from osteopenic individuals (Parfitt et al. 1983).

The problem of age-related bone loss may be compounded in females after the menopause since the oestrogen lack that follows ovarian failure leads to increased osteoclastic bone resorption during the fifth and sixth decades without an equivalent increase in bone formation. The administration of oestrogen to postmenopausal females can prevent this bone loss and subsequent increased risk of fracture since it inhibits osteoclastic bone resorption in these circumstances (Al-Azzawi et al. 1987).

Osteoporosis may also result from the effects of increased blood levels of glucocorticosteroids. These can be produced in excessive amounts by the adrenal glands as a result of a pituitary adenoma secreting excessive adrenocorticotrophic hormone (ACTH), by a tumour or hyperplasia of the adrenal glands or by the inappropriate secretion of ACTH by a malignant tumour, such as lung cancer. Occasionally, an ACTH-producing pituitary tumour may cause excessive pressure-related expansion of the sella turcica in the base of the skull. When associated with concomitant osteoporosis in an ancient skeleton, this finding might suggest that there had been excessive steroid production during life. However, one cannot rely upon these two findings to confidently make such a diagnosis since, on the one hand, most pituitary tumours that cause expansion of the pituitary fossa do not produce ACTH, and on the other hand, osteoporosis is a common finding in elderly subjects.

The precise mechanism of action of corticosteroids in bone resorption is unclear. Steroid-related reduced calcium absorption from the gut may cause secondary hyperparathyroidism but, unlike most cases of primary or secondary hyperparathyroidism, there appears to be suppression of bone formation or at least a significant reduction in the coupling of formation to sites of previous bone resorption. Osteoporosis may also occur in patients with a variety of inflammatory diseases due to the side-effects of steroids given therapeutically. These include asthma, rheumatoid arthritis, and other autoimmune and systemic inflammatory diseases, such as systemic lupus erythematosus. Osteoporosis can also occur in patients who have excessive production or administration of thyroid hormones since these cause increased bone turnover with resorption slightly exceeding formation.

3 Cytokine-Related Bone loss

Cytokines such as interleukin-1 (IL-1), tumour necrosis factor (TNF), and interleukin-6 (IL-6) are released in high concentration at sites of acute and chronic inflammation by inflammatory cells, such as leucocytes and macrophages. IL-1 and TNF are powerful stimulators of bone resorption in vitro and in vivo (Mundy 1988) and they are considered to be involved in the bone loss that occurs around joints in rheumatoid arthritis and at sites of infection in bone. They may also be released locally in bone by tumour cells or by tumour-related immune cells to account, in part at least, for the osteolysis seen around metastatic deposits in bone.

3.1 Rheumatoid Arthritis

Rheumatoid arthritis is an autoimmune disease that affects many tissues including the synovial membrane of joint capsules. The synovial villi become enlarged due to oedema and infiltration by leucocytes. These cells release a variety of factors and enzymes which cause pain and swelling of the joint capsule as well as destruction of the articular cartilage and bone. Almost any joint can be affected but the metacarpo/phalangeal joints, knees, wrists, and shoulders are involved most frequently. Permanent damage to the metacarpo/phalangeal joints and joint capsules causes characteristic ulnar deviation of the fingers. The local release of osteoclast activating factors into joints is likely to be responsible for the osteopenia that is seen around affected joints. Systemic administration of cytokines such as IL-1, TNF, and lymphotoxin cause systemic osteoclastic resorption. Thus, the generalized osteoporosis that is seen in some patients with rheumatoid arthritis may be due, in part at least, to the systemic effects of cytokines on bone resorption.

3.2 Osteoarthritis

Osteoarthritis is more common than rheumatoid arthritis and is caused by excessive wear and tear on joints coupled with a poorly understood susceptibility of joint cartilage to degradative enzymatic action in some individuals. In the early stages of the disease there is fibrillation of the surface of the joint cartilage which gradually becomes eroded down to the underlying bone. There is often proliferation of new bone at the joint margins in the later stages of the disease and this may limit joint movement. In response to the joint damage, synovial villi may become enlarged and infiltrated with inflammatory cells and thus can appear similar to those seen in rheumatoid arthritis. Release of cytokines by these inflammatory cells could aggravate existing damage. The knees, hips, and spine are affected most frequently and fusion of vertebral bodies can lead to severe limitation of cervical and lumbar movement.

3.3 Periodontal Disease

Periodontitis is a common dental disorder that is characterized by the presence of infection and inflammation of gums. The local release of osteoclast activating factors by immune cells can lead to loss of bone around affected teeth. Inflammation typically causes localized rather than generalized bone loss from the jaws.

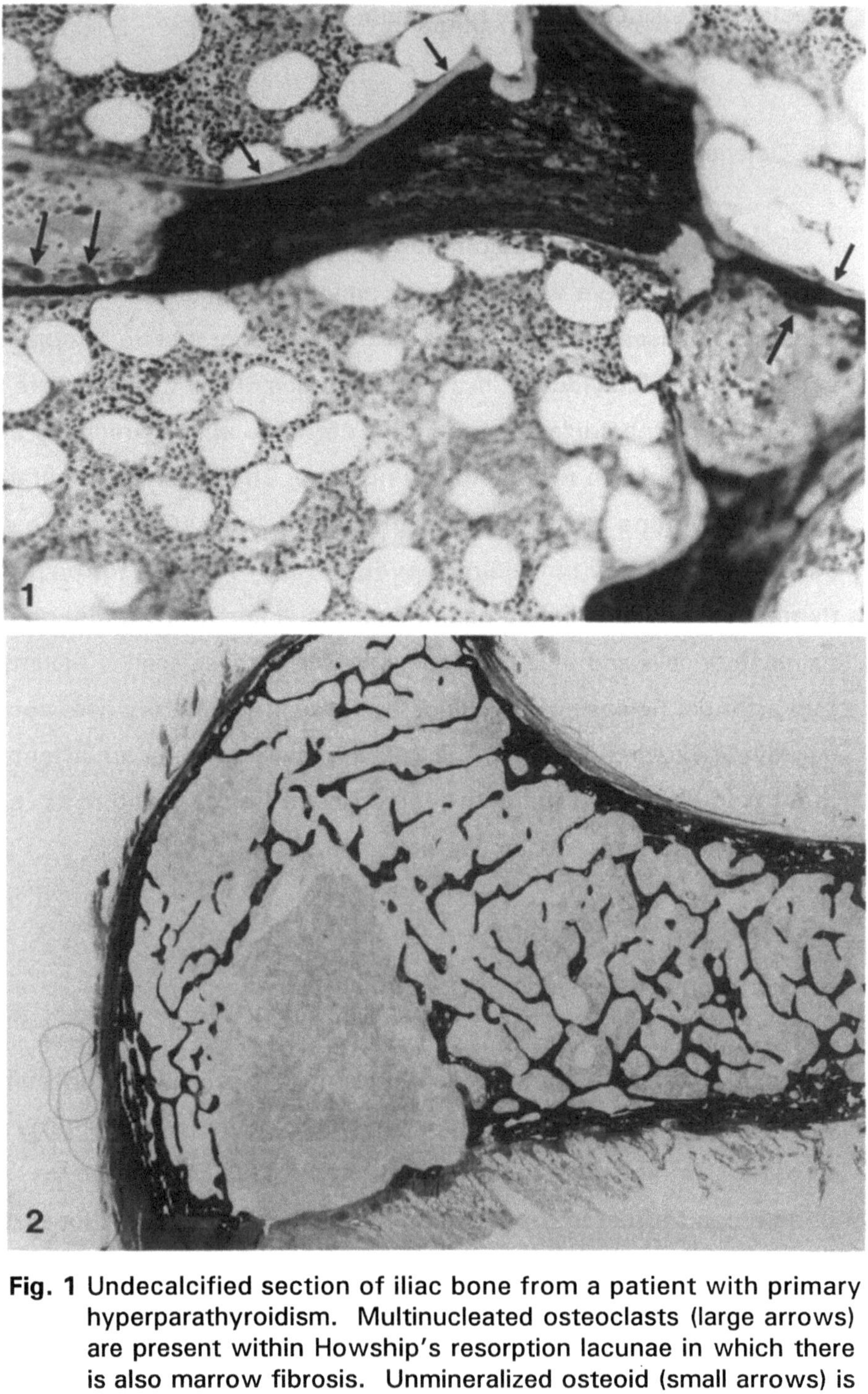

Fig. 1 Undecalcified section of iliac bone from a patient with primary hyperparathyroidism. Multinucleated osteoclasts (large arrows) are present within Howship's resorption lacunae in which there is also marrow fibrosis. Unmineralized osteoid (small arrows) is present or part of the bone surface. (1% aqueous toluidine blue, original magnification x165)

4 Malignancy

Bone is a common site to which malignant tumours metastasize. Most tumour deposits, for example, lung and breast cancer, cause osteolysis (Fig. 2) while some, for example, prostatic cancer, are typically osteosclerotic. Malignant tumour cells can secrete a number of osteoclast activating factors such as growth factors and cytokines (Mundy 1998). These factors cause the local proliferation of osteoclasts from bone marrow precursors and these, rather than the malignant cells, directly resorb the bone. Some malignancies, for example multiple myeloma, may cause the appearance of multiple lytic lesions within bones due to the local release of cytokines such as lymphotoxin, IL-1 and IL-6. Malignant tumours can also affect bone turnover without metastasizing by secreting osteoclast activating factors into the blood stream. These factors can then cause a generalized increase in osteoclastic resorption and thus be responsible for humorally mediated hypercalcaemia of malignancy (Mundy 1988). Some, such as parathyroid hormone-related protein, may also have a PTH-like effect on the kidney to increase calcium reabsorption and so contribute to the hypercalcaemia.

Tumour cells can also secrete osteoblast activating factors, most of which have yet to be fully characterized and purified. A list of known osteoblast stimulating factors is given in Table 2. Some of the osteoclast activating factors such as IL-1, TNF and platelet-derived growth factor (PDGF) have also been shown to stimulate indices of bone formation in some in vitro organ culture systems (Canalis,

Fig. 2 Undecalcified section of iliac bone from a patient with metastatic breast cancer. Two osteolytic deposits of metastatic tumour are present near the iliac crest. (1% aqueous toluidine blue, original magnification x8)

McCarthy and Centrella 1988). Thus, it is possible that some factors may stimulate resorption and/or formation and that their net effect might be determined by, for example, the local concentration and duration of release. Prostatic cancer is the best characterized osteosclerotic metastatic tumour and recent studies have indicated that an osteoblast stimulating factor(s) can be secreted by hyperplastic and neoplastic as well as normal postpubertal but not prepubertal prostatic cells (Canalis et al. 1988).

TABLE 2 Osteoblast-stimulating Factors

Insulin like growth factors I and II (IGF)
Parathyroid hormone, insulin
Vitamin D metabolites
Anabolic steroids, thyroxine
Bone morphogenetic protein (BMP)
Prostaglandins
TGFb, IL-1, TNF, macrophage-derived factors

The radiological appearances of metastatic tumours in bone depend upon the number of metastatic deposits and whether they are osteolytic or osteosclerotic. Metastatic tumours are much more common in adults than primary bone tumours which tend to arise in one site in bone and metastasize to soft tissues.

4.1 Paget's Disease

Paget's disease of bone is not a metabolic bone disease since calcium and/or hormonal imbalance is not the primary underlying disorder. It is a monostotic or, more typically, polyostotic disorder caused by excessive bone resorption by larger than normal osteoclasts at affected sites. This is followed by excessive new bone formation, much of which is woven and laid down in irregularly shaped resorption lacunae giving rise to the typical mosaic pattern seen histologically in the sclerotic bone. In the early stages of the disease, lytic defects may be seen radiologically at affected sites and these later become sclerotic. Despite the increased mass, the bone is structurally weaker than normal and there is an increased incidence of fracture and deformity which can lead to arthritis in the nearby joints. Malignant tumours develop in affected bones in less than 1% of individuals with this condition.

5 Summary

The process of normal bone remodelling can be disturbed in metabolic bone diseases due to an increase in the number of microscopic sites of bone turnover. This may lead to a reduction in bone mass if there is an imbalance between the amount of bone removed by osteoclasts and the amount laid down subsequently by osteoblasts. There are many known osteoclast activating factors which are secreted locally or systemically in excessive amounts in malignancy and in metabolic bone and joint diseases to account for the increased bone loss that can be observed in affected bones. At this time, there are fewer known osteoblast activating factors and these have been less well characterized. However, the list of known agents is growing. The local release of such factors could account for the increased bone mass seen around some secondary tumour deposits and inflammatory lesions in bone.

References

Al-Azzawi F, Hart DM, Lindsay R (1987) Long-term effect of oestrogen replacement therapy on bone mass as measured by dual photon absorptiometry. Brit Med J 294:1261-1262

Boyce BF, Fell GS, Elder HY (1985) Hypercalcaemic osteomalacia due to aluminium toxicity. Lancet ii:1009-1013

Canalis E, McCarthy T, Centrella M (1988) Growth factors and the regulation of bone remodelling. J Clin Invest 81:277-281

Chambers TJ (1985) The pathobiology of the osteoclast. J Clin Pathol 38:241-252

Ellis HA, Pearl KM (1973) Azotaemic renal osteodystrophy: a quantitative study on iliac bone. J Clin Pathol 26:83-101

Koutsilieris M, Rabbini SA, Bennet HPJ, Goltzman D (1987) Characteristics of prostate-derived growth factors for cells of the osteoblast phenotype. J Clin Invest 80:941-946

Mundy GR (1988) Hypercalcemia of malignancy revisited. J Clin Invest 82:1-6

Parfitt AM (1986) The cellular basis of bone remodelling: the quantum concept reexamined in light of recent advances in the cell biology of bone. Calcif Tissue Int 36:537-545

Parfitt AM, Mathews CHE, Villaneuva AR (1983) Relationships between surface, volume and thickness of iliac trabecular bone in ageing and in osteoporosis: implications for the microanatomic and cellular mechanism of bone loss. J Clin Invest 72:1396-1409

Verbeuken AH, van de Vyver FL, van Grieken RE, de Broe ME (1985) Microanalysis in biology and medicine: ultrastructural localization of aluminum. Clin Nephrol 24:s58-s77

Initial Stages of Systemic Bone Disease

Michael Schultz
Zentrum Anatomie
Georg-August-Universität
Kreuzbergring 36
W-3400 Göttingen
Germany

Palaeopathology is a discipline that plays a significant part in the reconstruction of the way of life of prehistoric and historic populations (Schultz 1982). In recent years, microscopic investigation has become more important in the diagnosis of ancient diseases. Often, the result of a macroscopic or radiological investigation is not able to provide a reliable diagnosis. Light microscopy and scanning electron microscopy are good tools with which to confirm a diagnosis (Schultz 1988a). In many cases microscopy can detect vestiges of diseases, not visible by gross examination. Thus, initial stages of bone diseases can also be clarified by microscope techniques. Nevertheless, some authors still do not think much of palaeohistopathology: "...as a general rule, in archeological skeletal material, microscopic data add little to what can be seen grossly or on X-ray films" (Ortner and Putschar 1981, p. 52).

On the other hand, there are authors with long experience in the histopathology of dry bones (e.g. Hackett 1976). Some of them have published results, demonstrating that particular diseases cause a typical lesion in the bone matrix, which cannot be mistaken for the lesion of a different disease. For example, Hackett (1976), in his classical article on treponematosis, published some figures which described very clearly the different stages of such a specific disease. The basis for this convincing elucidation of the stages lay in long years of experience in histopathology. When studying the microstructure of dry bones, it is striking that the bony tissue does not always react in the same way, when affected by an inflammation (Schultz and Teschler-Nicola 1987a). This fact is well demonstrated in the case of bone tissue damaged by treponematosis (Weber 1927c Hackett 1976; Schultz and Teschler-Nicola 1987a).

However, the basis of microscopic research relies not only on experience, but also on a sufficiently good technique in producing thin or ground-sections of dry bone samples. There are several methods described on how to obtain adequate specimens (see Schultz 1988a).

It is well known that post mortem soil erosion often causes changes which could be mistaken for the product of bone disease (Wells 1967; Schultz 1986a). For these alterations Wells (1967) created the expression "pseudopathology". As the bones of neonates and infants are very fine and fragile, they can be severely affected by all kinds of decomposition. In these cases, it is often very difficult to differentiate intra vitam changes, especially the initial stages, from post mortem alterations. Only a few typical examples will be presented here in order to demonstrate the efficiency of histodiagnosis.

1 Porotic Structures of the Skull Vault of Infant Skeletons

Occasionally, the external surface of infant skulls show slightly porotic structures (examples are also shown in Schultz, 1990a, fig. 2 and Schultz 1990b, Table 14;2 and Table 15;2). These changes could be caused by post mortem erosion (decomposition) or by a pathological process. If the lesion is caused by a pathologic process, anaemia, rickets, or inflammation (e.g. ostitis, osteomyelitis) it could be considered for differential diagnosis. It is exceptionally difficult to make a reliable diagnosis by macroscopic study alone, if the porotic structures are only slightly developed and if the diseased skull bone is part of a neonate or young infant skeleton. Thus, in these cases, light microscopy is absolutely necessary. In the case of a fully developed porotic structure, e.g. caused by anaemia, diagnosis is easier. As is well known, the skull bone is built up of external and internal laminae, and the spongy bone called diploe (Fig. 1).

In the case of anaemia (stage 1) the internal lamina is not destroyed and shows a normal structure (Fig. 2); also the diploe is usually not affected. Occasionally, the external section of the diploe shows slightly enlarged cavities. The external lamina is usually partly disintegrated (Fig. 2). In the next stage (stage 2) the skull bone becomes thicker. As the bone mainly grows in the area of the external diploe, the cavities are clearly enlarged. The internal lamina is not involved, while the external lamina is almost completely destroyed (Fig. 3). In the last stage (stage 3) the affected skull vault has thickened markedly. The trabeculae of the diploe take on a parallel orientation, i.e. the diploic trabeculae are arranged radially with thinning of the external lamina (Fig. 4). In X-rays, the typical "hair-on-end phenomenon" has been established (Steinbock 1976). In the region of the thickened skull vault, the external lamina is no longer observable (Fig. 4). Macroscopically the skull shows the signs of spongy hyperostosis (see, for example, El-Najjar et al. 1976).

In the case of rickets, pathological alterations are mainly found on the external skull surface. Very small squamous appositions cover the external lamina and give the bone a porotic aspect. As they partly grow together they may become a confluent layer. Occasionally, both anaemia and rickets can affect an infant skull. In this case the correct diagnosis is extremely difficult (for examples see Schultz 1990b, Tale 14;3; Schultz 1984).

In the case of inflammation of the skull bones (i.e. ostitis and osteomyelitis) various kinds of bone structures can be observed. Porotic structures, especially at the beginning of the disease, are frequently seen. They look very similar to the defects caused by anaemia (for examples see Schultz 1987, Fig. 6; Schultz 1990b, Table 15;2). Often, though not as a rule, the vestiges of inflammation of the skull bones in neonates and young infants are associated with vestiges of an inflammation of the periosteum (i.e. periostitis). In such cases, the child died before typical vestiges of a periostal process could develop on the bone surface. Since the variety of newly formed bone structures caused by osteomyelitis or ostitis is large, a typical case is described.

The skull vault of a 6- to 12-month-old child from the early Bronze Age cemetery of Ikiztepe (Anatolia) shows only a slight porotic structure on the external lamina, while the internal lamina is characterized by a partly porotic, partly squamous layer (see also Schultz 198, Fig. 6; Schultz 1990b, Table 15.2,3). In the thin section, the porotic bone surface can be diagnosed as the product of hypervascularization. The small holes are the foramina of blood vessel canals (not microfistulae !). Thus, the structures are quite different to the structures seen in the hypertrophy of the diploe in anaemia. Here, the holes represent bone marrow cavities. The light microscope investigation of the partly porotic, partly squamous layer on the internal lamina detects a multilayered structure (Fig. 5). The examination by polarized light shows that the organization of the layer had already begun. The stratification of the layer and the state of organization demonstrate that the process was underway before the child died. Furthermore, the microscopic investigations show that there was also an affection of the dura mater (suspicion of pachymeningitis). This kind of destruction of the diploe and the evidence of repair seen as a metaplastic process (Fig. 5) verify the reaction described above. These findings and the results of the examination of the postcranial skeleton yield a clear diagnosis: non-specific haematogenic osteomyelitis.

The importance of microscopic investigation for the epidemiology of diseases in prehistoric populations can be demonstrated with an example from the early Bronze Age population from Ikiztepe in northern Anatolia.

Out of a total of 144 infant burials, 129 skeletons with evidence of anaemia, rickets or osteomyelitis and 117 skeletons with evidence of irritation of the meninges were suitable for skull examination (Schultz 1990b). The first investigation was carried out at the site of excavation using only macroscopic techniques. The frequency of the four diseases was as follows:

Anaemia	7.0 %
Rickets	0.0 %
Osteomyelitis	8.5 %
Irritation of meninges	6.8 %

After microscopic examination, according to the new diagnosis, the frequency of the four diseases was as follows:

Anaemia	4.7 %
Rickets	3.9 %
Osteomyelitis	4.7 %
Irritation of meninges	9.5 %

The differences in the frequency of all four diseases is striking. Three cases were macroscopically diagnosed as osteomyelitis but microscopically diagnosed as irritation of meninges. Three other cases were macroscopically diagnosed as anaemia but microscopically diagnosed as osteomyelitis. Five cases were macroscopically diagnosed as osteomyelitis but microscopically diagnosed as rickets.

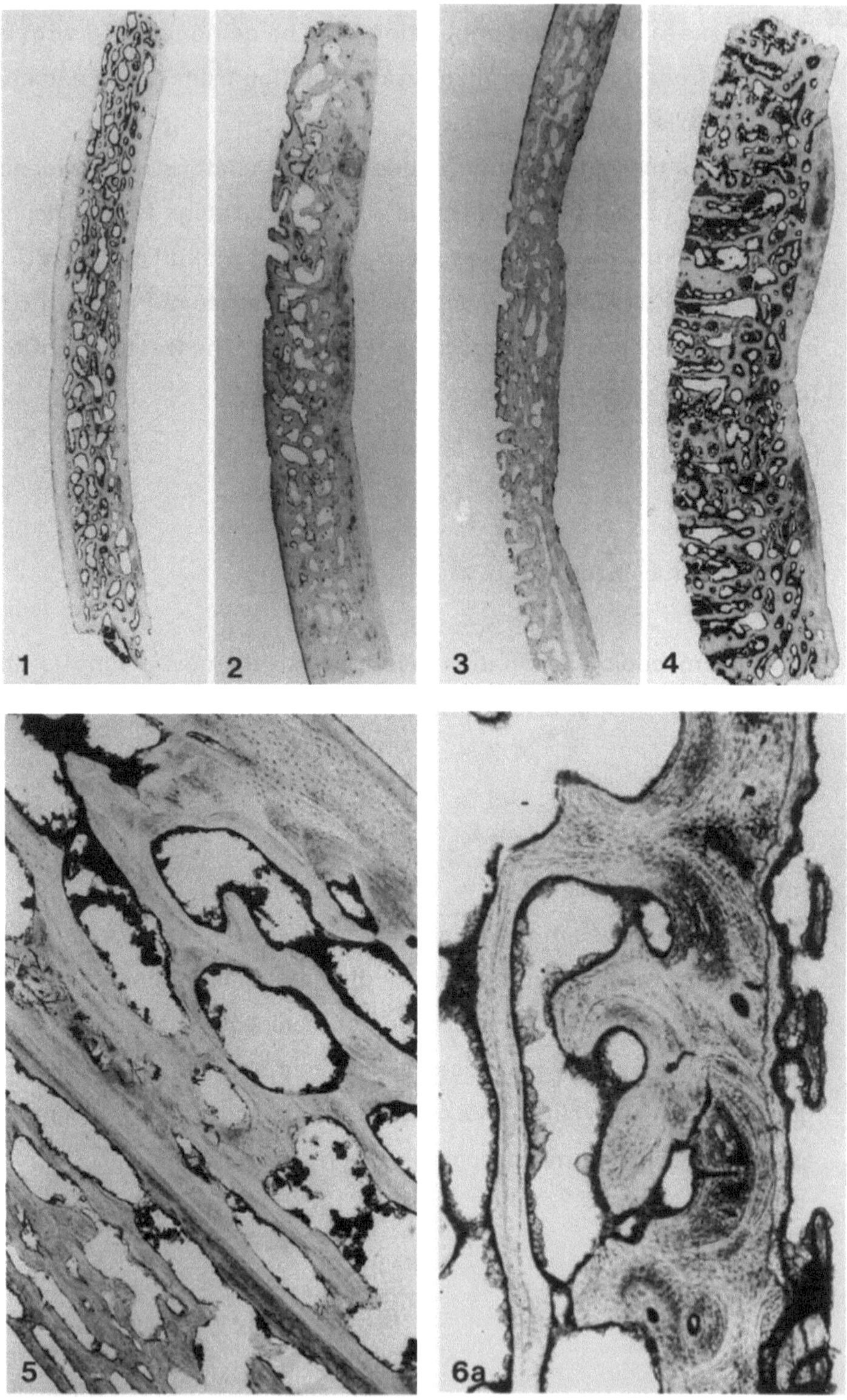
1
2
3
4
5
6a

Fig. 1 Thin section through the skull vault of a 5-year-old child from the neolithic site of Alburg-Straubing in lower Bavaria (Germany); normal bone structure

Fig. 2 Thin section through the skull vault of a 1 1/2- to 2- year-old child from the early Bronze Age cemetery of Ikiztepe in northern Anatolia (Turkey); anaemia (stage 1)

Fig. 3 Thin section through the skull vault of a 2- to 3-year-old child from the early Bronze Age cemetery of Ikiztepe in northern Anatolia (Turkey); anaemia (stage 2)

Fig. 4 Thin section through the skull vault of the child presented in Fig. 1; anaemia (stage 3)

Fig. 5 Thin section through a fragment of the parietal bone of a 6- to 12-month-old child from the early Bronze Age cemetery of Ikiztepe in northern Anatolia (Turkey). The multilayered, newly formed bone structure is seen below the internal lamina. The diploe shows signs of disintegration and metaplastic processes. Osteomyelitis; magnification x25

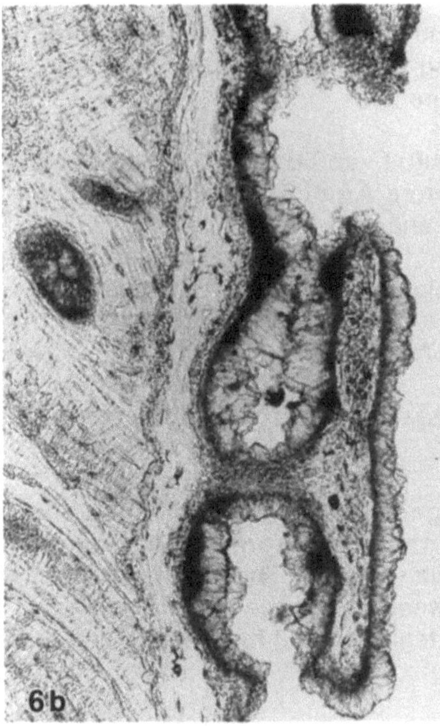
6 b

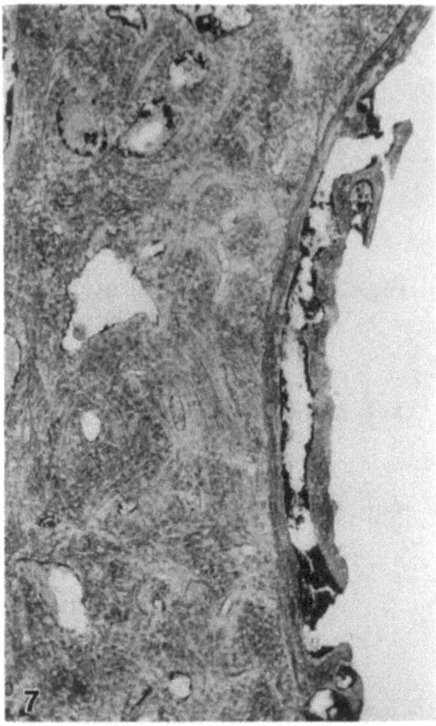
7

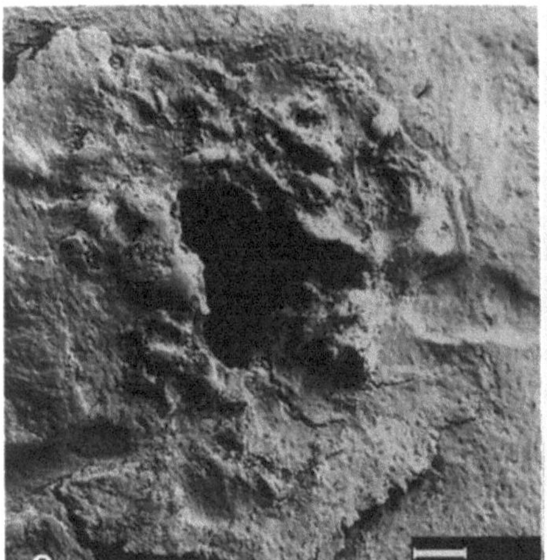
8
200 µm

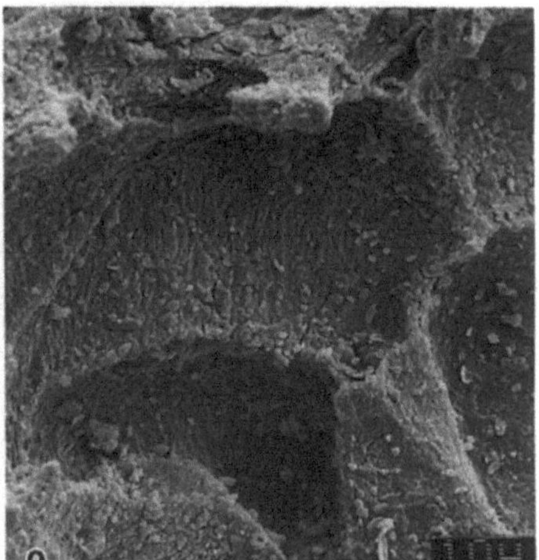
9

Fig. 6 Thin section through the frontal bone of a 2- to 3-year- old child from the early Bronze Age cemetery of Ikiztepe in northern Anatolia (Turkey). As the product of an inflammatory process of the meninges (meningitis, pachymeningitis) small, calcified, scale-like particles, connected to the internal lamina by pedicles, are seen. The bone surface is covered by small crystals of post mortem origin;
a magnification x25;
b magnification x100

Fig. 7 Thin section through the frontal bone of a 5- to 7-year-old child from the early Byzantine cemetery of Bogazköy in central Anatolia (Turkey). Calcified, plaque- like structure on the internal lamina can be seen. Haemorrhagic meningitis; magnification x25

Fig. 8 Tumour metastasis in the temporal bone of an adult individual from the Roman cemetery of Linz in upper Austria (Austria). View of the external lamina; the newly developed metastasis has grown around a small blood vessel canal. Scanning electron microscopy; magnification: see scale

Fig. 9 Howship's lacunae in a malignant osteoclastic tumour in the skull base of a late adult male from the Coptic cemetery of Sayala in Nubia (Egypt). Scanning electron miroscopy; magnification: see scale

2 Porotic Structures on the Orbital Roof

Porotic structures on the orbital roof, known as cribra orbitalia, are frequently seen in prehistoric and historic infant skulls. They can occur in three different stages (Schultz 1988b): slight (stage 1), middle (stage 2), and severe (stage 3).

The causes of their occurrence is usually restricted to anaemia (e.g. Stuart-Macadam 1985; 1987) or possibly to racial or geographic factors (cf. Henschen 1961). The results of recent microscopic investigations, however, have shown that cribra orbitalia were, in many cases, caused by inflammation in the adjoining region (Schultz 1987; Götz 1988), such as the paranasal sinuses (e.g. sinusitis frontalis), the endocranial cavity (e.g. pachymeningitis), or the skull vault (e.g. osteomyelitis). Microscopic examinations of cribra orbitalia detect in the case of anaemia enlarged, bubble-like marrow cavities in the diploe and, in the case of inflammation irregular foci of liquefaction and a dilation of the bone trabeculae (Schultz 1987; Götz 1988). In cases of anaemia, the bone trabeculae, however, could very often have a parallel orientation (see Schultz 1987, Fig. 28). Furthermore, in inflammatory processes small blood vessel impressions and signs of periostal reaction could be seen. Thus, the results of the light microscope and scanning electron microscope investigations described by Götz (1988) are a very important contribution to the aetiology of cribra orbitalia. Porotic structures in the orbital roof have also been observed in cases of rickets (Schultz 1987).

According to the results of Götz (1988), the aetiology of inflammation in the skull region is now much more comprehensible. This leads us not only to a better understanding of the resistance of prehistoric people to infection, but also gives us more knowledge on the correlation between some external environmental factors (Schultz 1982).

3 Alterations on the Internal Lamina of the Skull Vault

Vestiges of irritations of the meninges (e.g. inflammation, trauma, epidural haematoma, subdural abscess, tumour) on the internal lamina of the skull bones are very frequent in prehistoric and historic skeletal material (Schultz 1987). However, as a rule, the signs are so very slight that they are usually neglected by the palaeopathologist.

Initial stages of this group of diseases can be easily detected by a magnifying glass or an endoscope. The most frequent disease here seems to be inflammation of the meninges i.e. meningitis or pachymeningitis (Schultz 1987, 1991). As is well known, the meningitis can be associated with increased cerebrospinal pressure and/or haemorrhage and/or small-grained, scale-like soft tissue formations which may calcify secondarily. The vestiges of all these structures can be differentiated in prehistoric and historic skulls (for examples see Schultz 1984, 1987, 1990a, b, 1991). Occasionally, the increasing cerebrospinal pressure can cause a hydrocephalus to form (Schultz 1987; Schultz and Teschler-Nicola 1987b).

As an example of a typical case of meningitis/pachymeningitis, we have taken the case of a 2- to 3- year-old child from the early Bronze Age cemetery of Ikiztepe in northern Anatolia. The internal lamina shows the signs of slightly increased pressure of the cerebrospinal fluid and the impressions ofthe cerebral gyri are deeper than normal. At the bottom of these digital impressions numerous small, scale-like particles are found; this is a sign of inflammation (cf. photographs of the macroscopic view in Schultz 1987, Fig. 18). The microscopic examination proves that these small, brittle particles are of intra vitam origin. They are fastened by their bases by short pedicles onto the internal lamina of the skull bone (Fig. 6). As the microstructure (e.g. collagen) of the bone is clearly observable, it is easily seen by polarized light microscopy that these particles are not ossified, but only calcified (see Schultz 1987, Figs. 18, 19). Thus, these particles represent the calcified product of an inflammation, which was originally of soft tissue character.

The following case is typical for a small haematoma formation in the meningeal area (Fig. 7). The internal lamina of the skull vault of a 5- to 7-year-old child from the early Byzantine cemetery in Bogazköy in central Anatolia shows a thin, plaque-like porotic structure. Upon examination of the thin section with the microscope, it is seen that the thin, irregular structure is localized close to the internal lamina of the skull bone. The examination by polarized light microscopy shows that, at the time of death, the plaque-like structure was only calcified and not ossified. The localization and the microstructure confirm that it is the product of a haemorrhagic and inflammatory process of the meninges, apparently a slight bleeding in the region of a small vein running into the sagittal sinus, associated with (pachy)meningitis.

Several stages of partly organized, typical epidural haematoma are described by photographs in other articles (Schultz 1986b, 1987, 1990a,b). Recent studies (Schultz 1987) have shown that, if the patient survived the bleeding, the organization of an epidural haematoma is strongly characterized by the growth of very small, net-like blood vessels. These blood vessels form typical impressions in the internal lamina (pressure atrophy). Type and development of these blood vessel impressions demonstrate the stage of organization of such an epidural haematoma (see also photographs in Schultz 1987). As a rule, the blood vessel impressions are reduced or may even disappear after the complete healing of the lesion.

4 Alterations on the External Surface of the Diaphysis of Long Bones

The health standards of prehistoric populations are often represented by the state of health of their children. This means that the physical condition of the juveniles is a reliable measure for the health condition of a particular population. In many cases chronic malnutrition induces restriction of the immunological system. The resulting condition is an increase in the frequency of inflammations, which may also affect the osseous system. This reflection may explain why the frequency of chronic malnutrition and osteomyelitis in neonate and infant skeletons of prehistoric and, occasionally, of historic populations show almost the same relative levels (Schultz 1991).

Unfortunately, the macroscopic features of the diaphysis of a baby or a young infant affected by chronic malnutrition (e.g. scurvy) look very similar to the diaphysis of a child suffering from an early inflammation (e.g. periostitis). In both cases the shaft of the long bone may be covered by a secondary layer consisting of calcified or even ossified tissue (Schultz 1986a; Schultz and Teschler-Nicola 1987a).

The histological examination of the layer demonstrates, in its microstructure and kind of attachment (Schultz 1986a,b; 1988a), whether the new bone growth is caused by, for example (1) an organized and calcified haematoma (induced by trauma under scurvy); (2) an involucrum (osteomyelitis); or (3) a periostal reaction (periostosis or periostitis, see Schultz 1986a). Sometimes, it is possible to differentiate microscopically between periostosis and periostitis (Schultz 1986a; Schultz and Teschler-Nicola, 1987a). Such layers are not always found in a generalized, i.e. systemic, state but may also appear in a circumscribed state.

Initial stages of inflammatory bone diseases, e.g. osteomyelitis, are very difficult to diagnose, especially in the long bones of babies or young infants. A relatively positive sign for an inflammatory process on bone surfaces is the existence of an increased number of blood vessel canals representing hypervascularization. Usually, the foramina of these canals are relatively enlarged.

Vestiges of osteoclastic activity and irregular, bulky bone trabeculae in the marrow cavities and/or the metaphyses of the long bones, without external, i.e. cortical, or periostal reaction is extremely rare (Schultz 1984, 1986a). In addition to X-ray investigations, the microscope examination of transverse thin sections can help to achieve a reliable diagnosis.

5 Initial Stages of Tumourous Diseases

Vestiges of tumourous lesions are relatively frequent in prehistoric skeletal material (Brothwell 1967; Strouhal 1976; Steinbock 1976, Uhlig 1982; Schultz 1986a). Therefore, it is doubtful whether tumours should be characterized as a so-called civilization disease (Schultz 1982).

Many tumours diagnosed in the skeleton do not belong to the group built up of fully developed tissue. Therefore, they usually grow very rapidly. Superficial lesions, caused by this tumour group, are normally marked by the great number of blood vessel canals. These newly developed blood vessels often represent the focus of an early metastasis ("mini-metastasis", Fig. 8). Thus, the external bone surface, affected by a tumour, may have an extremely porotic appearance. It is not rare that a bone affected by an early stage of osteomyelitis (see above) sometimes looks very similar to a tumourous bone. Consequently, it is not easy to make a reliable diagnosis. Incidentally, it should not be forgotten that soil erosion (decomposition) may also cause similar porous alterations in bones (see above). These short remarks show that macroscopic inspection is not satisfactory for the provision of correct information on the cause of disease.

It is well known that there are osteoblastic as well as osteoclastic tumours. Several typical examples are published in the literature (e.g. Uhlig 1982; Grupe 1988; Schultz 1986a). For osteoblastic tumours irregular, sometimes bulky, sometimes fine bone trabeculae are characteristic, while in osteoclastic tumours the numerous Howship's lacunae are microscopically the characteristic structure (Fig. 9). In many tumours, both vestiges of osteoblastic and osteoclastic reaction can be found, though as a rule, one kind of reaction predominates.

In many tumours, especially in the group of solid, osteoblastic neoplasms, regularly structured osteons are missing (Schultz 1978, 1987). In the place of regular Haversian systems, "Faserfilz-Osteone" (Knese et al. 1954; Schultz 1978) are principally found. This kind of osteon can also be regarded as an initial symptom of an osteoblastic process in compact bone tissue (Schultz 1986a). For differential diagnosis it is important to know that "Faserfilz-Osteone" can, occasionally, also be seen in normally structured bone tissue (e.g. in the mandibular corpus, in the bony attachment of muscles).

When prehistoric bone findings are examined under the microscope, it is easily seen that in most cases tumorous lesions show a pattern typical for their structure, growth and development. Using these results in combination with the results produced by the macroscopic and radiological examination, diagnosis is exact.

6 Summary

This short communication describes typical lesions of initial stages of bone diseases. Furthermore, the possibilities and the importance of histodiagnosis are demonstrated. As examples, five groups of bone changes characterizing special morphological features are presented: (1) porotic structures on the skull vault, caused by anaemia, rickets, osteomylitis, ostitis, and periostitis; (2) porotic structures on the orbital roof (so-called cribra orbitalia), caused by anaemia, rickets, and inflammatory processes in the adjoining region (e.g. paranasal sinuses, endocranial cavity or skull vault); (3) alterations on the internal lamina of the skull vault, caused by meningitis, pachymeningitis, and epidural haematoma; (4) alterations on the external surface of the diaphyses of long bones, caused by scurvy, osteomyelitis, ostitis, and periostitis; and (5) initial stages of tumorous diseases (e.g. osteoblastic and osteoclastic lesions).

Acknowledgements. This study was supported by the Deutsche Forschungsgemeinschaft (Schu 396/3-3, Schu 396/4-3, Schu 396/7-1).

References

Brothwell DR (1967) The evidence of neoplasms. In: Brothwell DR, Sandison AT (eds) Diseases in antiquity. Thomas, Springfield, pp 320-345

El-Najjar MY, Ryan DJ, Turner CG, Lozoff B (1976) The etiology of porotic hyperostosis among the prehistoric Anasazi Indians of southwestern United States. Am J Phys Anthropol 44:477-488

Götz W (1988) Histologische Untersuchungen an Cribra orbitalia. Ein Beitrag zur Paläopathologie des Orbitadaches. Med Diss, Univ Göttingen

Grupe G (1988) Metastasizing carcinoma in a medieval skeleton: differential diagnosis and etiology. Am J Phys Anthropol 75:369-374

Hackett CJ (1976) Diagnostic criteria of syphilis, yaws and treponarid (treponematoses) and some other diseases in dry bones. Sitzungsber Heidelberger Akad Wiss 4:351-470

Henschen F (1961) Cribra cranii, a skull condition said to be of racial or geographic nature. Pathol Microbiol 24:724-729

Knese K-H, Voges D, Ritschl I (1954) Untersuchungen über die Osteon- und Lamellenformen im Extremitätenskelett des Erwachsenen. Z Zellforsch 40:323-360

Ortner DJ, Putschar WGJ (1981) Identification of pathological conditions in human skeletal remains. Smithsonian Contrib Anthropology 8

Schultz M (1978) Ein Knochentumor an einem präkolumbischen Indianerschädel aus Peru. Ein Beitrag zur Differentialdiagnose knöcherner Schädeltumoren. Virchows Arch A Pathol Anat Histol 378:121-132

Schultz M (1982) Umwelt und Krankheit des vor- und frühgeschichtlichen Menschen. In: Wendt H, Loacker N (eds) Kindlers Enzyklopädie. Der Mensch, vol 2. Kindler, Zürich, pp 259-312

Schultz M (1984) The diseases in a series of children's skeletons from Ikiztepe, Turkey. In: Capecchi V, Rabino Massa E (eds) 5th Eur Meet Paleopathology Association, Siena 1984. Univ Press, Siena, pp 321-323

Schultz M (1986a) Die mikroskopische Untersuchung prähistorischer Skelettfunde. Anwendung und Aussagemöglichkeiten der differentialdiagnostischen Untersuchung in der Paläopathologie. In: Amt für Museen und Archäologie BL, Liestal; and Anthropologische Forschungsinstitut Aesch (eds) Tagungsber Paläopathol Symp Liestal (Baselland) 21.-23. Jun 1984. Amt für Museen und Archäologie BL, Liestal/Schweiz. Ber Arbeit Amt Mus Archäol Kanton Baselland 6:1-140

Schultz M (1986b) Der Gesundheitszustand der frühmittelalterlichen Bevölkerung von Bogazkale/Hattusa. In: Kültür ve Turizm Bakanligi TC (eds) 8 Kazi, Arastirma ve Arkeometri Semp, 26-30 May 1986. Arastirma Sonuclari Toplantisi 4. Eski Eserler ve Müzeler Genel Müdürlügü, Ankara, pp 401-409

Schultz M (1987) Spuren unspezifischer Entzündungen an prähistorischen und historischen Schädeln. Ein Beitrag zur Paläopathologie. Med Habil, Univ Göttingen

Schultz M (1988a) Methoden der Licht- und Elektronenmikroskopie. In: Knussmann R (ed) Anthropologie. Handbuch der vergleichenden Biologie des Menschen, vil I/1. Fischer, Stuttgart, pp 698-730

Schultz M (1988b) Paläopathologische Diagnostik. In: Knussmann R (ed) Anthropologie. (Handbuch der vergleichenden Biologie des Menschen, vol.I/1. Fischer, Stuttgart, pp 480-496

Schultz M (1990a) Causes and frequency of diseases during early childhood in Bronze Age populations. In: Capasso L (ed) Advances in paleopathology. Proc 7th Eur Meet Paleopathology Association, Lyon, Sept 1988. Marino Solfanelli Editore, Chieti. J Paleopath, Monogr Publ 1: 175-179

Schultz M (1990b) Erkrankungen des Kindesalters bei der frühbronzezeitlichen Population vom Ikiztepe (Türkei). In: Andraschko FM, Teegen W-R (eds) Gedenkschrift für Jürgen Driehaus. Zabern, Mainz, pp 83-90

Schultz M (1991) Erkrankungen des Kindesalters bei der frühbronzezeitlichen Population von Hainburg/Niederösterreich. Mitt Anthropol Ges Wien (MAGW) 118/119:369-380

Schultz M, Teschler-Nicola M (1987a) Krankhafte Veränderungen an den Skeletten aus dem Karner der St. Martinskirche in Klosterneuburg, Niederösterreich. III. Entzündliche Veränderungen an den langen Röhrenknochen. Ann Naturhist Mus Wien 89:252-296

Schultz M, Teschler-Nicola M (1987b) Ein Hydrocephalus aus der mittelneolithischen Kreisgrabenanlage von Kamegg, Niederösterreich. Ein Beitrag zur Ätiologie kindlicher Schädelerkrankungen. Ann Naturhist Mus Wien 88:223-251

Steinbock RT (1976) Paleopathological diagnosis and interpretation. Bone diseases in ancient human populations. Thomas, Springfield, Ill

Strouhal E (1976) Tumors in the remains of ancient Egyptians. Am J Phys Anthropol 45:613-620

Stuart-Macadam PL (1985) Porotic hyperostosis: representative of a childhood condition. Am J Phys Anthropol 66:391-398

Stuart-Macadam PL (1987) Porotic hyperostosis: new evidence to support the anemia theory. Am J Phys Anthropol 74:521-526

Uhlig C (1982) Zur paläopathologischen Differentialdiagnose von Tumoren an Skeletteilen. Landesdenkmalamt Baden-Württemberg, Stuttgart. Materialh Vor Frühgesch Baden-Württemberg I

Weber M (1927) Schliffe von mazerierten Röhrenknochen und ihre Bedeutung für die Unterscheidung der Syphilis und Osteomyelitis von der Osteodystrophia fibrosa sowie für die Untersuchung fraglich syphilitischer, prähistorischer Knochen. Beitr pathol Anat Allg Pathol 78:441-511

Wells C (1967) Pseudopathology. In: Brothwell DR, Sandison AT (eds) Diseases in antiquity. Thomas, Springfield, Ill, pp 5-19

The Uses of Scanning Electron Microscopy in the Interpretation of Some Examples of Trauma in Human Skeletal Remains

Jennifer Wakely
Department of Anatomy and School of Archaeological Studies
University of Leicester
University Road
Leicester LE1 9HN
UK

For the purposes of this paper a wide interpretation is placed on the word 'trauma' to include all forms of damage inflicted on human bones by human activity. Soft tissue damage not involving bone will not, except in highly unusual circumstances, such as the accidental preservation of a body in a peat bog or deliberate preservation in a burial rite such as mummification, normally enter the archaeological record. This paper will consider examples of weapon injury, surgical intervention and burial ritual from a variety of sites.

Studies of this nature combine experimental studies on contemporary bone with examination of archaeological material. This is essential in order to establish firm diagnostic criteria for distinguishing between the effects of human activity and taphonomic changes, for example animal gnawing or the effects of weathering. One then uses the argument, proposed by Shipman (1981, 1988) that if a mark on an archaeological specimen shows the same macroscopic and microscopic features as marks deliberately induced in a contemporary experiment it is reasonable to assume that the 'old' and 'new' marks were produced in a similar manner, by purposeful, human activity.

Criteria used are, on the macroscopic level, the position of the marks and their appearance to the naked eye. For example, battlefield weapon injuries inflicted in formal, hand-to-hand fighting, tend to be localized on the left side or frontal region of the skull (Inglemark 1939; Manchester 1983; Wenham 1990). Butchering marks on animal bone, or marks on human bone inflicted in defleshing burial rituals are usually localized in areas of muscle attachment, whereas scrapes and grooves of taphonomic origin are randomly distributed over the bone surface (Andrews and Cook 1985; Cook 1986a,b; Fulchieri et al. 1986; Olsen and Shipman 1988; Shipman 1988). Cuts into living bone, or bone that is recently dead, with sharp-edged metal weapons, have flat, smooth, often polished surfaces that are clearly distinguishable from the rough surface produced by fracturing or cutting 'dry' bone from which the organic matrix has decayed (Brothwell 1971; Wenham 1990).

Scanning electron microscopy has the facility to produce high-resolution images, with three dimensional information, of the surface of bone specimens or replicas (Figs. 1-3). It has been used with success in a number of archaeological investigations (Olsen 1988), for example to identify cutting, chopping and scraping marks on bone marked by stone implements (Cook 1986b; Shipman 1981; Shipman and Rose 1983) and metal weapon injuries on Anglo-Saxon (Wakely and Wenham 1990) and Bronze Age skeletons (Wakely and Bruce 1989).

Large, fragile and valuable archaeological samples cannot, naturally, be placed directly in the scanning electron microscope. The standard technique is to prepare first a latex negative replica of the surface under investigation and then, from this, an epoxy resin positive replica, which can be gold-coated and viewed in the scanning electron microscope without damage to the specimen or the instrument (Shipman and Rose 1983; Cook 1986b; Bromage 1987). In the author's experience only intact, natural bone surfaces can be replicated. Surfaces which are fragile due to poor preservation may be damaged by the latex replication procedure. Specimens which have been coated with shellac or other varnishes make poor subjects because the surface topography may be filled in by the varnish. Surface coatings deposited as a result of cultural practices, for example, pigments, may also be lifted. It is important to consult an experienced conservator before attempting to cast a specimen about which any of these doubts exist. The remainder of this paper will describe some examples from personal experience in examining human bone trauma by SEM. All specimens were viewed in an ISI DS130 SEM operated at 15 kV, and photographed on Kodak Technical Pan film. Magnifications of between x10 and x300 yield the most useful information, though higher magnifications have been used on occasion.

1 Injuries Made by Sharp Metal Weapons

Figures 1-3, illustrate one of the injuries observed on an axis vertebra from a Bronze Age site at Covesea, Scotland. Seven, separate, flat, smooth cuts on the superior surface of this bone (Fig. 1) suggest that the individual had been decapitated, severing the spine between atlas and axis. Four of the cuts, including the one illustrated in Figs. 2 and 3, lie in the same plane, the other three are separate from the first four and from each other, suggesting a sequence of four separate blows. The facets on the vertebral body and arch show faint suggestions of gently curved parallel striations. They are visible when viewed by low-magnification light microscopy, and by SEM they can be seen at high resolution (compare Figs. 2 and 3).

Similar patterns have been observed on cut bone surfaces from sources as diverse as skeletons from an Anglo-Saxon cemetery in Eccles, Kent (Wenham 1990; Wakely and Wenham 1989), skulls from the Iron Age war cemetery at Maiden Castle, Dorset (Stevens and Wakely, unpublished) and a trophy head from Borneo (Beavitt and

Wakely, unpublished). They are clearly a general feature of bone that has been cleanly cut with a single blow from a thin, straight, metal blade. Experimental studies (Fig. 4; Wenham 1990) in which contemporary bone was cut with a replica of an Anglo-Saxon sword produced a similar pattern of parallel striations on the bone and periosteum that could be attributed to irregularities, some too small to be seen except by SEM, in the edge of the blade. Very fine striations, visible at magnifications of around 1200x, can be seen in well preserved archaeological material. Above this magnification the texture of the bone itself obscures the details of the injury.

One interesting feature of the Covesea axis is that the striations on the four linked facets are curved so that they form part of a single wide arc, such as would be made by one swinging a long-bladed weapon, eg. a sword, with some force across the bone. Their direction relative to the anatomy of the bone as a whole suggests that the assailant was standing behind and to the right of the victim.

While cuts and scrapes made by stone and metal implements have some features in common (Shipman and Rose 1983; Olsen 1988) the finer striations and greater likelihood of "chattermarks" if metal is used could possibly, but not certainly, indicate the use of metal. However, SEM images of trauma from bronze and iron weapons are indistinguishable. Unfortunately, therefore, this method cannot be used to indicate the transition from bronze to iron weapons.

2 Surgical Wounds

It is usually only possible to detect evidence of surgery archaeologically if it leaves traces on the skeleton, so, for example, trephination is the oldest 'surgical' procedure described in the archaeological record, and one for which the SEM is proving a valuable interpretative method. It is possible to discern some aspects of the operator's technique and, where there was post-operative survival, examine the process of healing. Interpretation of ancient trephinations draws on comparison with experimental cuts made on bone with stone implements, since many ethnographic and archaeological examples of this procedure come from non metal-using cultures. Such cuts have several major distinguishing features. They are V- or U-shaped in section with tapered ends. They bear on their sides fine parallel striations caused by minor irregularities in the blade edge. They show asymmetry with one rough side, or "shoulder" and one smooth side (Shipman 1981; Shipman and Rose 1983; Bromage and Boyde 1984). More controversially, Bromage and Boyde (1984) attribute the asymmetrical appearance and the slope of the flakes on the rough side to 'handedness', ie. that the rough side of a cutmark always lies on the side nearest to the operator's little finger, because of the tendency to supinate the hand when cutting. More experimental work is needed, using a large sample of operators of left and right hand preferences, to establish whether a general principle does in fact exist here.

Five methods of trephination have been described in the archaeological record. The use of a specialized instrument, the trephine, to cut out a round section of bone was uncommon in antiquity but became widely used in the mediaeval and post-mediaeval periods (Lisowski 1967; Parker et al. 1986; Brothwell 1974). Sawing, in which four straight cuts are used to remove a rectangle of bone, and drilling are risky because of the poor control they afford over entry through the inner table of the skull. The two methods most commonly described ethnographically and in European archaeological material are grooving and scraping (Brothwell 1981; Lisowski 1967; Parker et al. 1986). Both these methods give a slow, controlled entry into the cranial cavity. In grooving the operator cuts round a circular or subcircular track, gradually removing a roundel of bone and producing a relatively steeply bevelled opening. In scraping the same area of bone is repeatedly pared until an opening appears in the inner table of the skull. This technique produces an elongated opening with widely sloping sides. To some extent "grooving" and "scraping" are not mutually exclusive terms: the sides of a groove will, of course, have been scraped, and the centre of a scraped opening will have much in common with a groove, particularly if the operator changes the angle of approach to deepen the opening.

From the SEM point of view trephinations are of great interest. The manner in which they are produced should reveal itself in the pattern of marks made by the operator's instrument in and around the opening. This should provide reliable diagnostic criteria for differentiating the genuine trephination from the many other openings, often collectively called "pseudotrephinations" (Janssens 1987) that occur in ancient skulls, caused, for example, by weapon injury, rodent gnawing or even excavation damage. The key point is that in scraping, or grooving, the operator repeatedly cuts at the same area of bone but is unlikely to cut twice in exactly the same place. Intersecting cutmarks, as opposed to the single set of parallel lines caused by a weapon blow, would be predicted. A limited sample of European skulls have been examined by SEM, by Wakely and Duhig (1990).

The Beaker period skull from Crichel Down, Dorset, England has a large opening, 75 x 51 mm, on the left side. Its steeply bevelled sides point to grooving as the probable method (Piggott 1940). On the sides SEM shows a number of short, straight or widely curved, grooves, extending from the outer table, across the diploe and sometimes as far as the inner table (Fig. 5). Many of the grooves show evidence

Figs 1-3 An axis vertebra from Covesea, Scotland showing six cuts to the superior surface (1-6). Cuts 1, 4, 5 and 6 represent a single blow from a sharp metal weapon. Figs. 2 and 3 are views of cut 4 by light microscopy and SEM, to show (a) the parallel arrangement and curvature of the striations on the cut surface and (b) the greater resolution produced by SEM

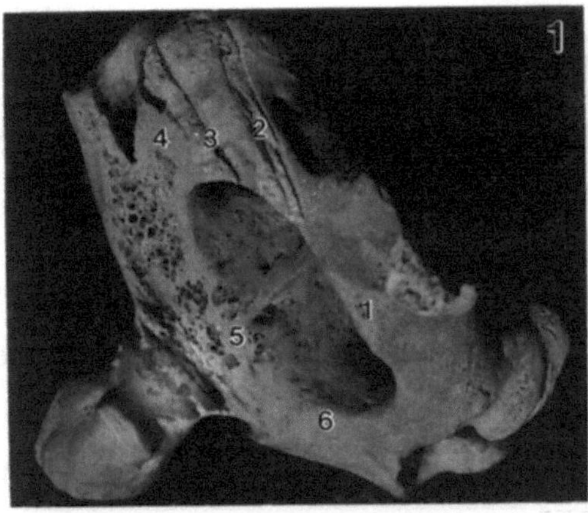

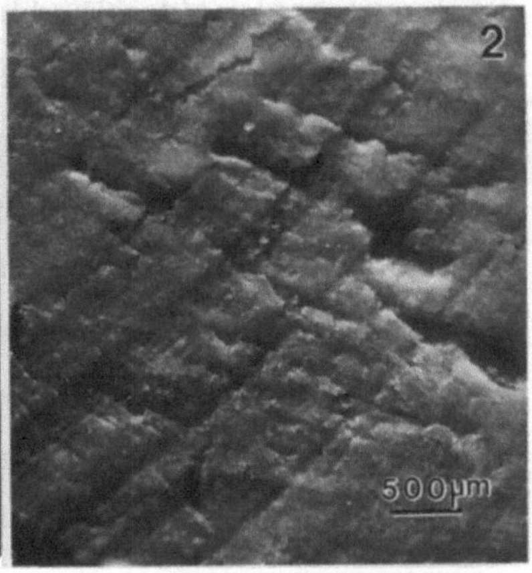

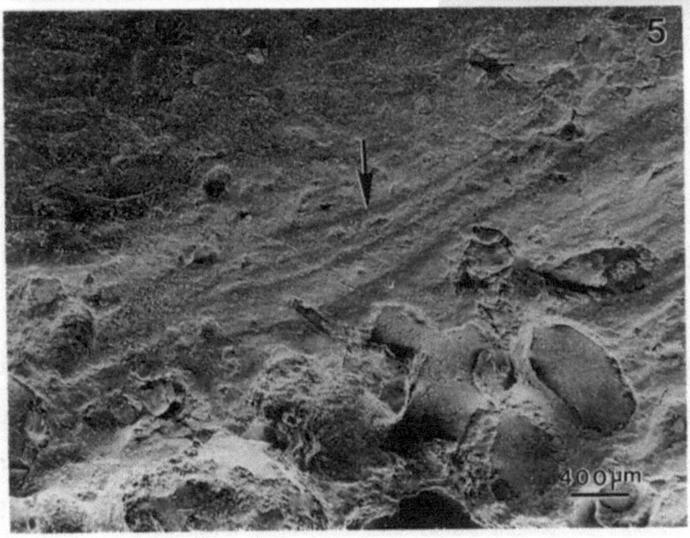

Fig. 4 Cut made on contemporary bone with a replica Anglo-Saxon sword, showing parallel striations on the cut surface

Fig. 5 Grooves on the side of the opening in the trephined skull from Crichel Down (arrowed)

of the asymmetry identified by Shipman and Rose as a feature of the marks created by deliberate cutting with a stone implement. This feature lay consistently towards the outer table of the skull, suggesting a consistent cutting direction.

An unprovenanced skull from Girton, Cambridge, possibly Romano-British or Anglo-Saxon, presents a D-shaped opening with widely bevelled sides, characteristic of trephination by scraping, in the midline of the skull vault immediately behind the bregma. The sides of the opening and a surrounding triangle of the outer surface of the skull bear many intersecting striations (Fig. 6). Some individual groups of striations descend from the outer table surface on to the sides of the opening. Others are cut off abruptly by striations running in a different direction, as the operator changed the scraping direction.

In spite of the hazards associated with opening the skull without the benefit of modern surgical techniques there is archaeological and ethnographical evidence of recovery. Figure 7 shows part of the margin of a healed opening from a Neolithic skull from Bougon Deux Sèvres, France. The margins of the opening are rounded, not sharp. There are no striations and new bone, with vascular channels visible by SEM was forming at the time of the individual's death.

3 Burial Practices

Secondary burial is a rite in which the body is excarnated by natural weathering, scavenging animals or deliberate cleaning, leaving only the bones to be interred. It is widespread in Neolithic Europe, and found in many parts of the world even in the present (Atkinson 1965; Ucko 1969; Grinsell 1975; Tainter 1978; Burl 1981; Bloch and Parry 1982; Hedges 1983; Henderson 1988; Gilkes 1989).

Purposeful removal of flesh, whether human or animal, will leave cutmarks concentrated in areas of muscular or ligamentous attachment to bone. Such patterned cutmarks have been observed on human bones from the glacial period from Gough's Cave, Somerset, England by Cook (1986b). Also from the Palaeolithic are examples described by Ullrich (1982) and Russell (1987). De Laet (1958) and Fulchieri et al. (1986) describe bones from Neolithic burials in which cutmark distribution suggests an intentional removal of soft tissue.

The single specimen examined by the author is part of a humerus from a Neolithic burial in Haddenham, Cambridgeshire, England. Two groups of cutmarks are visible, one in the area of attachment of the brachialis muscle, the other in the attachment of the medial head of the triceps. The cuts are parallel and sharply defined (Fig. 8). By SEM they show the V-shaped profile, tapering ends and internal striations characteristic of cuts made with a stone implement (Fig. 9). These features, taken together with the correlation with muscle attachment, suggest that this bone is an example of excarnation. The seven individuals represented by the bones found in the burial in question were poorly preserved and no other cutmarks were found. The question of cannibalism is often raised in connection with cut human bone (see, for example, Ullrich 1982), but probably cannot be answered from skeletal evidence alone.

4 Summary

Scanning electron microscopy is an additional technique, complementary to the examination of sections, that can be used to investigate the fine structure of archaeological human bone. Replica methods render it non-destructive to the actual specimen. Its facility for producing high-resolution images of surfaces has been used to describe and interpret some examples of intentional damage to human bone, in the course of such culturally important activities as conflict, medicine and rituals associated with the disposal of the dead.

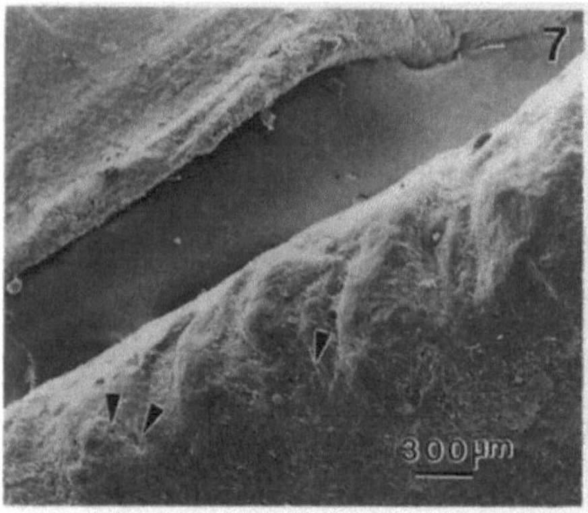

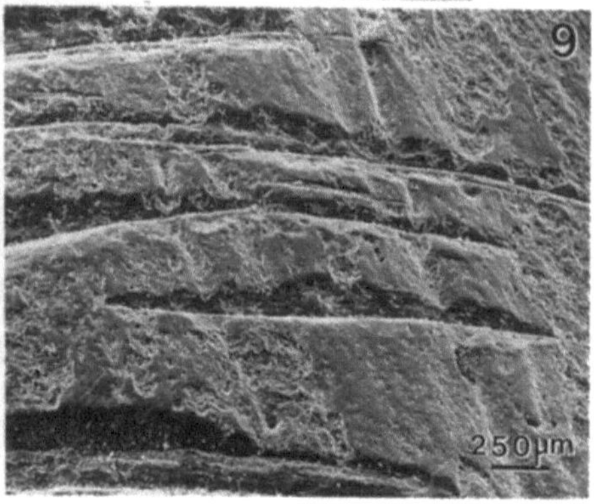

Fig. 6 Part of the opening in the skull from Girton and surrounding area of skull vault, showing superimposed and intersecting striations

Fig. 7 Part of the opening of a healed trephination in the Neolithic skull from Bougon Deux Sèvres. Note the rounded undulating surface. The holes (arrowed) are evidence of vascular channels

Figs. 8 and 9
A group of parallel cuts on the medial side of the humerus from Haddenham, viewed by SEM in **b**

Acknowledgements. It is a pleasure to thank the following colleagues for access to specimens and laboratory facilities: Dr. M.F.Bruce, University of Aberdeen, for the axis vertebra from Covesea; Mr. P.Beavitt, University of Leicester, for the trophy skull from Borneo; Dr. R.Foley, Dr. J.Garlick and Ms. C.Duhig, University of Cambridge, for skulls from Maiden Castle, Crichel Down, Bougon Deux Sèvres and Girton, all in the Duckworth Collection, Cambridge; Dr. S.J.Wenham, for experimental sword-cut on bone, Dr. K.Manchester and Dr. C.A.Roberts, University of Bradford, for Anglo-Saxon skulls from Eccles loaned to Dr. Wenham for her undergraduate project in the Anatomy Department, University of Leicester; Dr. I.Hodder, Cambridge and Ms. F.Lee, Bradford, for the humerus from the excavation at Haddenham.

I am also pleased to thank Mr. G.L.C.McTurk of the Leicester University Scanning Electron Microscope Unit for operation of the ISI DS130 SEM, Mrs. J.d'Lacey for typing the manuscript, Ms. D.Miles for assistance with photography, Dr.D.J.Pallot for the facilities of the Anatomy Department, University of Leicester, and Her Majesty's Inspector of Anatomy in connection with the experimental studies.

References

Andrews P, Cook J (1985) Natural modifications to bone in a temperate setting. Man 20:675-691

Atkinson RJC (1965) Wayland's Smithy. Antiquity 39:126-133

Bloch M, Parry J (1982) Death and the regeneration of life. Univ Press, Cambridge

Bromage TG (1987) The scanning electron microscopy/replica technique and recent applications to the study of fossil bone. Scann Electron Microsc 1(2):607-613

Bromage TG, Boyde A (1984) Microscopic criteria for the directionality of cutmarks on bone. Am J Phys Anthrop 65:359-366

Brothwell DR (1971) Forensic aspects of the so-called skeleton Q1 from Maiden Castle, Dorset. World Archaeol 3:233-241

Brothwell DR (1974) Osteological evidence for the use of a surgical modiolus in a Romano-British population. J Arch Sci 1:209-211

Burl A (1981) Rites of the gods. Dent, London

Cook J (1986a) The application of scanning electron microscopy to taphonomic and archaeological problems. In: Roe DA (ed) Studies in the Upper Palaeolithic of Britain and northwest Europe. BAR Int Ser 296:143-163

Cook J (1986b) Marked human bones from Gough's Cave, Somerset. Proc Univ Bristol Spelaeol Soc 17(3):275-285

De Laet SJ (1958) The low countries. Thames & Hudson, London

Fulchieri E, Rabino Massa E, Doro Garetto T (1986) Differential diagnosis between palaeopathological and non-pathological post-mortem environmental factors in ancient human remains. J Human Evol 15:71-75

Gilkes JA (1989) Cave burials of northern England. Brit Archaeol 11(1):11-15

Grinsell LV (1975) Barrow, pyramid and Tomb. Thames & Hudson, London

Hedges JW (1983) Isbister, a chambered tomb in Orkney. BAR Brit Ser 115:214-221; 266-271

Henderson J (1988) Factors determining the state of preservation of human remains. In: Boddington A, Garland AN, Janaway RC (eds) Death, decay and reconstruction. Univ Press, Manchester, pp 43-54

Inglemark BE (1939) The skeletons. In: Thordemann B (ed) Arms and armour from the Battle of Wisby, 1361. Stockholm, pp 149- 209

Janssens PA (1987) Etude paléopathologique de neuf cranes de l'abbaye cistercienne de Coxyde (Belgique), traumatismes vrais et faux, la pseudotréphination. Helinium 27:109-127

Lisowski FP (1967) Prehistoric and early historic trephination. In: Brothwell DR, Sandison AT (eds) Diseases in antiquity. Thomas, Springfield Ill, pp 651-672

Manchester K (1983) The archaeology of disease. Univ Press, Brad ford, pp 62-63

Olsen SL (1988) The identification of stone and metal tool marks on bone artefacts. In: Olsen SL (ed) Scanning electron microscopy in Archaeology. BAR Brit Ser 452:337-360

Olsen SL, Shipman P (1988) Surface modification on bone: trampling versus butchery. J Arch Sci 15:535-553

Parker S, Roberts C, Manchester K (1986) A review of British trepanations with a report on two new cases. OSSA 12:141-157

Piggott S (1940) A trepanned skull of the Beaker period from Dorset and the practice of trepanning in prehistoric Europe. Proc Prehist Soc 3:112-132

Russell MD (1987) Mortuary practices at the Krapina Neanderthal site. Am J Phys Anthropol 72:381-397

Shipman P (1981) Application of scanning electron microscopy to taphonomic problems. Ann NY Acad Sci 276:357-385

Shipman P (1988) Actualistic studies of animal resources and hominid activities. In: Olsen SL (ed) Scanning electron microscopy in Archaeology. BAR Brit Ser 452:261-285

Shipman P, Rose J (1983) Early hominid hunting, butchering and carcass-processing activities: approaches to the fossil record. J Anthropol Archaeol 2:57-98

Tainter JA (1978) Mortuary practices and the study of prehistoric social systems. Adv Arch Meth Theor 1:105-141

Ucko PJ (1969) Ethnography and archaeological interpretation of funerary remains. World Archaeol 1:262-280

Ullrich H (1982) Artificial injuries on fossil human bones and the problem of cannibalism, skull-cult and burial rites. Anthropos (Brno) 21:253-262

Wakely J, Bruce MF (1989) Interpreting signs of trauma on a human axis vertebra. J Anat 167:265

Wakely J, Duhig C (1990) A comparative microscopical study of three European trephined skulls. J Palaeopathol 3:75-87

Wakely J, Wenham SJ (1989) Microscopical features of blade injuries to bone surfaces in six Anglo-Saxon skeletons from Eccles, Kent. In: Roberts CA, Lee F, Bintliff J (eds) Burial archaeology. Current research, methods and developments. BAR Brit Ser 211:291

Wenham SJ (1990) Anatomical interpretations of Anglo-Saxon weapon injuries. In: Hawkes SJ (ed) Weapons and warfare in Anglo-Saxon England. Oxford Univ Committee Archaeol Monogr 21:123-139

Index

MIX
Papier aus verantwortungsvollen Quellen
Paper from responsible sources
FSC® C105338

If you have any concerns about our products,
you can contact us on
ProductSafety@springernature.com

In case Publisher is established outside the EU,
the EU authorized representative is:
Springer Nature Customer Service Center GmbH
Europaplatz 3, 69115 Heidelberg, Germany

Printed by Libri Plureos GmbH
in Hamburg, Germany